Foundations of
Organic Chemistry

Ron B. Davis Jr., Ph.D.

PUBLISHED BY:

THE GREAT COURSES
Corporate Headquarters
4840 Westfields Boulevard, Suite 500
Chantilly, Virginia 20151-2299
Phone: 1-800-832-2412
Fax: 703-378-3819
www.thegreatcourses.com

Ron B. Davis Jr., Ph.D.

Visiting Assistant Professor of Chemistry
Georgetown University

Professor Ron B. Davis Jr. is a Visiting Assistant Professor of Chemistry at Georgetown University, where he has been teaching introductory organic chemistry laboratories since 2008. He earned his Ph.D. in Chemistry from The Pennsylvania State University, where his research focused on the fundamental forces governing the interactions of proteins with small organic molecules. After several years as a pharmaceutical research and development chemist, he returned to academia to teach chemistry at the undergraduate level.

Professor Davis's research has been published in such scholarly journals as *Proteins* and *Biochemistry* and has been presented at the Annual Symposium of The Protein Society. He also maintains an educational YouTube channel and provides interviews and content to various media outlets, including the Discovery Channel.

At Penn State, Professor Davis was the recipient of a Dalalian Fellowship and the Dan Waugh Teaching Award. He is also a member of the Division of Chemical Education of the American Chemical Society. ∎

Table of Contents

Table of Contents

Foundations of Organic Chemistry

Scope:

Chemistry is defined as the study of matter and its properties. With regard to this definition, the roots of the study of chemistry can be traced back to more than one ancient civilization. Most notably, the Greeks and Chinese each independently postulated thousands of years ago that there must be a small number of elemental substances from which all other things were created as admixtures. Remarkably, both civilizations theorized that air, earth, water, and fire were among those elements. It was much more recently, however—just about 300 years ago—that famed French nobleman and chemist Antoine Lavoisier correctly identified one of the elements experimentally. Lavoisier's discovery is often cited as the event that heralded the birth of chemistry as a proper science. Theorizing based on observation of natural systems began to give way to controlled testing of the properties of matter, leading to an explosion of understanding, the echoes of which are still ringing in modern-day laboratories.

Organic chemistry is the subject dedicated to the study of a deceptively simple set of molecules—those based on carbon. Even today, centuries after the most basic governing principles of this subject were discovered, many students struggle to make sense of this science. At the university level, professors are often in a race against time to dispense the vast body of knowledge on organic chemistry to their students before semester's end, leaving little time for discussion of exactly how this information came to be known or of just how new experimentation might change the world we live in. This course endeavors to fill that gap.

As humanity's understanding of chemistry grew, so did the library of elements that had been isolated and identified, yet even as this library of elements grew, one of the simplest of them—carbon—seemed to play a very special and indispensable role in many small molecules. This was particularly true of the molecules harvested from living organisms. So obvious was the importance of this role that chemists dubbed the study of the fundamental molecules of life "organic chemistry," a science that today has been expanded to include any molecule relying principally on carbon atoms as its backbone.

In this course, you will investigate the role of carbon in organic molecules—sometimes acting as a reactive site on molecules, sometimes influencing reactive sites on molecules, but always providing structural support for an ever-growing library of both naturally occurring and man-made compounds.

Other elements will join the story, bonding with carbon scaffolds to create compounds with a stunningly broad array of properties. Most notable are the elements hydrogen, nitrogen, oxygen, chlorine, and bromine. The presence of these elements and others in organic chemistry spices up the party, but none of them can replace carbon in its central role.

The goal of this course is to take the uninitiated student on a tour of the development and application of the discipline of organic chemistry, noting some of the most famous minds to dedicate themselves to this science in the past few centuries, such as Dmitry Mendeleev (of periodic table fame), Friedrich Wöhler (the father of modern organic chemistry), and Alfred Nobel (the inventor of dynamite and founder of the most influential scientific prize in the history of humanity). You will also meet some very famous scientists from other fields whose forays into organic chemistry helped shape the science, such as Louis Pasteur of microbiology fame and Michael Faraday, the father of electromagnetism.

Approximately the first half of the course is dedicated to building the foundations of understanding modern organic chemistry. In this portion of the course, you will investigate the structure of the atom, the energetic rationale for chemical bonding between atoms to create compounds, how specific collections of atoms bonded in specific ways create motifs called functional groups, and ultimately the ways in which the bonds in these functional groups form and break in chemical reactions that can be used to convert one compound into another.

Next, you will apply that understanding of organic fundamentals to more complex, but often misunderstood, molecular systems, such as starches, proteins, DNA, and more. In the final portion of the course, you will turn your attention to how organic chemists purify and characterize their new creations in the laboratory, investigating techniques as ancient as distillation and as modern as nuclear magnetic studies.

After completing this course, the successful student will have all of the tools needed to have a meaningful dialogue with a practicing organic chemist about the theory behind his or her work, the interpretation of the results that he or she obtains in the lab, and—most of all—the impact that modern experimentation in organic chemistry might have on the future of humanity. ■

Conjugation and the Diels-Alder Reaction
Lecture 19

In this lecture, you will be introduced to the phenomenon of conjugation, in which multiple pi bonds in resonance with one another lend extra stability to a compound or ion. You will learn a few structural arguments for why this added stability exists, and you will discover a new way of modeling the energies of conjugated pi systems in alkenes. In addition, you will scratch the surface of conjugated diene reactivity by exploring two reactions. Finally, you will examine a reaction conceived by Otto Diels and Kurt Alder that allows us to quickly and easily form two carbon-carbon bonds.

Conjugation
- The simplest conjugated compound possible in organic chemistry is the hydrocarbon 1,3-butene. Compounds with pi bonds, or the potential to have them, tend to have resonance contributors that can stabilize those compounds; 1,3-butadiene is no different, having resonance contributors that contain pi electron density on the intervening carbon-carbon bond.

- When this resonance involves the commingling of multiple pi bonds, as it does in this case, we refer to this special type of resonance as conjugation. Conjugation in a compound comes with some important structural, physical, and chemical properties.

- The requirement for conjugation is that double bonds are alternating. If the double bonds are more than one carbon-carbon bond apart, then an intervening sp^3 carbon prevents the pi electrons from exchanging by resonance.

Naming Conjugated Compounds

- In spite of what we now know about the remarkable mobility of pi bonds in conjugated compounds, when we try to name these compounds, we revert to the most stable resonance contributor. This is always the contributor with all neutral atoms and alternating pi bonds. The name is constructed in much the same way as simple alkenes, but with the alteration that before the "-ene" suffix we insert a prefix "di-," "tri-," "tetra-," etc.

- For example, what is commonly called vinylethylene would be named 1,3-butadiene. Creating ever-larger linear conjugated alkenes produces 1,3,5-hexatriene, and then 1,3,5,7-octatetraene, etc. This analogy extends to cyclic conjugated compounds. For example, 1,3-cyclohexadiene would be a conjugated compound as well.

Stability of Conjugated Species

- Just as we did with isolated alkenes, we can also survey the effect of conjugation on stability using heats of hydrogenation. Let's do this using a simple system with only a few isomers, such as pentadienes.

- As a reference, the heat of hydrogenation of 1-pentene is 125 kilojoules per mole. If we add a second double bond that is isolated from the other by multiple carbon-carbon bonds, we create 1,4-pentadiene, which contains two isolated double bonds. Not surprisingly, the heat of hydrogenation doubles because we have essentially added a second bond of exactly the same type.

- But hydrogenation of its regioisomer 1,3-pentadiene releases less energy than the sum of that for two isolated double bonds. It is this observation that first led to our understanding of the stabilizing effect of resonance.

- This stabilizing effect extends far beyond simple diene hydrocarbons. Resonance has similar effects on the stability of compounds containing heteroatoms, such as enols, carbocations, and many more.

Reactivity of Conjugated Alkenes

- Although conjugated systems are somewhat more stable than similar but isolated pi systems, they can and do undergo many of the addition reactions that we discussed previously for simple, isolated alkenes—with a few important caveats.

- For example, the highly connected pi systems of conjugated compounds like alkenes mean that they can often undergo more complex addition reactions with unexpected regioisomers. The classic example of this is the addition of a hydrogen halide to 1,3-butadiene.

- Markovnikov's rule states that the addition of hydrogen halides to alkenes will occur with the halide going to the more-substituted carbon. But what happens when that alkene is conjugated?

- In the reaction between 1,3-butadiene and hydrochloric acid, the pi electrons still attack the acidic hydrogen of hydrochloric acid, producing a secondary carbocation—but the presence of an adjacent *p* orbital means that this carbocation is resonance stabilized, placing some of the positive charge density much farther away from the newly acquired proton than we are accustomed to seeing.

- So, in the second step of our reaction, the chloride nucleophile now has a *choice* of where to react. The two potential results are the products of a 1,2 addition, which produces a terminal alkene in accordance with Markovnikov's rule, and a 1,4 addition, which produces an internal alkene, the likes of which Zaitsev would be likely to predict.

- So, in a single system, we have a microcosm of this classic feud going on. The real winner is probably Zaitsev. The more stable product is, in fact, the internal alkene. But the catch is that Markovnikov's product forms faster, because the nucleophilic attack by chloride in the second step has a lower activation energy barrier at the 2 position than at the 4 position.

- So, if we run this reaction at very low temperatures for very short periods of time, we get more of the 1,2-addition product. This is because the competing process, the 1,4 addition, happens very slowly and doesn't get a chance to take place. We call reactions run under conditions like this kinetically controlled reactions, and the corresponding product is called the kinetic product.

- But if we give the system plenty of heat and lots of time, giving the 1,4-addition a chance to take place, then it will be the major product, because both pathways are accessible to the reagents, giving them a choice. We call reactions run like this, with long run times and high temperatures, thermodynamically controlled reactions.

Reactions with long run times and high temperatures are called thermodynamically controlled reactions.

The Diels-Alder Reaction

- In the early decades of the 1900s, synthetic organic chemistry was earning recognition as a powerful field of study with tremendous potential for useful application in medicine, industry, and beyond. But as the 20th century dawned, the library of synthetic techniques available to organic chemists was still lacking.

- A particular challenge facing chemists was the need for techniques to create new carbon-carbon bonds. In 1928, Otto Diels and Kurt Alder devised a reaction scheme that exploited the geometry of conjugated diene molecular orbitals to produce not just one new carbon-carbon bond, but two in a single step.

- The Diels-Alder reaction is actually one of a class of reactions known as pericyclic reactions, which consist of the exchange of multiple pi electrons simultaneously between two molecules to create new sigma bonds between them. The Diels-Alder is designated a [4+2]-pericyclic reaction because it involves such a reaction between a conjugated diene (having 4 pi electrons) and one pi bond from another reagent known as a dienophile (having 2 pi electrons).

- The simplest example that meets these criteria is the reaction between *s-cis*-1,3-butadiene and ethene. To envision how this reaction takes place, let's re-create the molecular orbitals of both molecules. First, 1,3-butadiene has four molecular orbitals, of which the pi 2 is the highest occupied molecular orbital (HOMO) and the pi 3 is the lowest unfilled molecular orbital (LUMO).

- We collectively refer to the HOMO and LUMO as frontier molecular orbitals, because they usually represent the orbitals that will be vacated and occupied, respectively, when a reaction takes place.

- Ethene is a bit simpler because it only has two *p* orbitals in its pi system. So, pi 1 is the HOMO and pi 2 is the LUMO.

- As the two reagents approach one another, the HOMO of the diene and the LUMO of the dienophile are in phase on both sides, and the same is true of the diene LUMO and the dienophile HOMO. This means that the two reagents are perfectly configured to exchange electrons on both sides in a concerted process that produces two new sigma bonds from four of the pi electrons.

- The entropic penalty of making one molecule from two is easily balanced and, in fact, overwhelmed by the enthalpic benefit of trading two pi bonds for two sigma bonds. The new carbon-carbon bond creation is complete.

- What makes the [4+2] reaction so easy to complete is the in-phase alignment of frontier molecular orbitals. Because the geometries of the molecular orbitals are perfectly set up for a reaction to take place, all we need is to mix the reagents and wait for the molecules to collide. We call this a thermally activated reaction.

- If we try to conduct analogous reactions—for example, in a [2+2] cyclization, making cyclobutene from two molecules of ethylene—the frontier molecular orbitals are not properly aligned when the compounds are in their lowest energy states.

- We can resort to a trick called photoactivation, promoting electrons into higher-energy molecular orbitals by shining light of just the right wavelength on them. This essentially changes the identity of the frontier molecular orbitals, artificially creating the overlap that we can get for free in a [4+2]. It is this detail—that Diels-Alder reactions are thermally activated—that makes them so useful.

- In addition, a number of biologically and commercially relevant compounds, such as cyclic terpenes, contain a cyclohexene motif. Compounds naturally occurring in plants, such as limonene in oranges, terpineol in lilac, and alpha Ionone from violet, can all be easily synthesized using a Diels-Alder strategy.

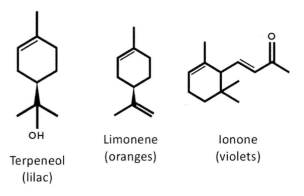

OH
Terpeneol
(lilac)

Limonene
(oranges)

Ionone
(violets)

- The Diels-Alder reaction can be performed with more complex dienes and dienophiles, leading to products with a more robust stereochemistry than the cyclohexene product from the previous example.

- In the addition of the dieneophile to the diene, both new carbon-carbon bonds form on the same side of the newly created ring. In the case of cyclohexene formation, this is purely an academic observation, because all substituents of the newly joined carbons are hydrogen atoms. But what if they are not?

- Let's begin by examining the first step of the Woodward synthesis of cholesterol. Robert Burns Woodward knew that he was trying to make a compound with four fused carbon rings on it from much simpler starting materials, so why not consider the Diels-Alder to start building this scaffold?

- Woodward began his synthesis using 1,3-butadiene to create his first fused ring to a compound known as hydroquinone. The Diels-Alder process ensured that the desired six-membered ring would form. But Woodard faced a new challenge: His bicyclic structure had the wrong stereochemistry. The Diels-Alder forced the addition of his new ring in a cis configuration.

- Consider the simpler example of 1,3-butadiene reacting with cis-2-butene as the dienophile. The product of the reaction will have chiral centers at the two carbons from the 2-butene. Because we used cis-2-butene, we expect to obtain a single meso compound in which the methyl substituents are cis to one another on the ring. (4R,5S)-4,5-dimethylcyclohexene.

- However, using trans-2-butene leads instead to a set of diasteromers in which the methyl groups are on opposite sides of the ring, creating an (R,R) enantiomer and an (S,S) enantiomer.

Berson, *Chemical Creativity*, Chap. 2.

McMurry, *Fundamentals of Organic Chemistry*, Chaps. 4.8.

Wade, *Organic Chemistry*, Chaps. 15.1–15.6, 15.11, 15.12.

Questions to Consider

1. For which of the reactions below do you expect different products to form depending on the temperature at which the reaction is run?
   a. 1,3-butadiene with 1 equivalent of HCl
   b. 1,4-pentadiene with 1 equivalent of HCl
   c. 1,3-cyclohexadiene with 1 equivalent of HCl

2. The Diels-Alder reaction between acrolein and 2-methoxy-1,3-butadiene can produce two different isomers. Which would you expect to form in excess, and why? (Hint: Consider charges in the resonance hybrid for each compound.)

Conjugation and the Diels-Alder Reaction
Lecture 19—Transcript

Steroid hormones are remarkably powerful organic molecules. They can go far beyond simply determining the gender of a developing fetus. I mean, just consider the market for the two most commonly known of these compounds—estrogen and testosterone. Both are produced and marketed as pharmaceutical products to fight medical conditions associated with aging.

Testosterone's ability to promote a process known anabolism, or the building up of complex organs and tissues in the body, has made it a useful substance in both the legitimate medical treatment of wasting diseases, and also as a performance enhancing drug of abuse in athletics. Estrogen and its derivatives are used as contraceptive medications. And they also, sometimes, are prescribed to post-menopausal women to help fight osteoporosis.

But it might surprise you to find out that both of these hormones, with their drastically differing metabolic effects, are made up from the same starting materials in the body. The common starting material in biological synthesis of these compounds is cholesterol. And, in the early part of the 1900s, a race was brewing to find a way to make this central compound in vitro. But one of the major challenges was finding ways to make new carbon-carbon bonds from smaller molecules.

Now, during this period in time, divining a new method to make this critical connection was practically a guaranteed Nobel Prize. In 1928, Otto Diels, a former student of the legendary Hermann Emil Fischer, and his own protege, Kurt Alter, published an article outlining a strategy then only called the diene synthesis. By the late 1940s, it was clear to all that this synthetic strategy was one of the keys which would ultimately unlock the total synthesis of cholesterol. In those days, a sort of holy grail of steroid chemistry.

In 1950, with the research groups of chemists Robert Robinson at Oxford and Robert Burns Woodward at Harvard independently closing in on this goal, Diels and Alder received their Nobel Prize for the diene synthesis. In honor of the tremendous impact their development has had on organic chemistry, we now call this reaction the Diels-Alder reaction. Hopefully, by the end of the day, you will see exactly how they did it, and why they so richly deserved this honor. But before we can begin to fully understand the Diels-Alder reaction, we need to take a step back and analyze a little bit simpler compound than these large hormone compounds. We need to take a look at compounds called conjugated dienes. Now, let's take a look at the simplest one right now.

Now, the very simplest conjugated compound possible in organic chemistry is this hydrocarbon called 1, 3-butene. Now, you may notice that the two double bonds of the compound appear to be separated by a carbon-carbon single bond, as I've drawn in here. Now, recall that compounds with pi bonds, or the potential to have them, tend to have resonance contributors, which can stabilize those compounds. And 1, 3-butadine. is no different, having resonance contributors which contain pi electron density in the intervening carbon-carbon bond.

Now, when this resonance involves the commingling of multiple pi bonds, as it does in our example here, we refer to this special type of resonance as conjugation. Conjugation in a compound comes with some important structural, physical, and chemical properties as well. You see, this is only achieved when all four of the p atomic orbitals align with one another. So, for example, if I try to rotate the central carbon-carbon bond here by about 90 degrees, I have a problem, because our two pi bonds will be oriented perpendicularly to one another, and because of that, they can't have pi electrons move from one to the other. In other words, I can't establish the resonance that I could previously.

Now, as you might suspect from this observation, conjugated dienes, like ours, have a strong tendency to take on that planar conformation, which aligns p orbitals of all the sp^2 carbon atoms in the compound. Now, all this leads to a potential pair of stereoisomers in our example. See, the partial pi bond character of that central carbon atom creates a restricted rotation, producing two different isomers.

Now, we call these S-cis, or cisoid, which looks like this. Or I can flip the central pi bond and make it an S-trans, or transoid, conformation. So the cisoid and transoid isomers, they tend to interconvert more quickly, because there's actually only a pi bond that needs to break. So a bona fide cis and trans alkene, while they can't interconvert well at all, would be interconverting much more slowly than this. Nonetheless, the two different states, cisoid and transoid, can exist as individuals.

Now, you can see from this that the requirement for conjugation is that the double bonds are alternating. If the double bonds are more than one carbon-carbon bond apart, let's try that. Let's put the double bonds more than one carbon-carbon bond apart. Well, now there's an intervening sp^3 carbon, and because the intervening sp^3 carbon has no available p orbitals, the electrons from the two pi bonds cannot exchange by resonance. And alternately, let's bring up a compound where the double bonds are not spaced at all; they're right next to one another in what we call accumulated diene. Now, in this case, notice that the central carbon has to have two different p orbitals participating in this bonding arrangement, and that means that it has to have two pi bonds which are not parallel, but rather, are normal to one another. So in neither of these two cases could we possibly make a compound that's conjugated. It's only when we have that very special arrangement of alternating pi bonds that we see this very special effect.

Now, in spite of what we know now about the remarkable mobility of pi bonds in conjugated compounds, when we try to name them, we revert to the most stable resonance contributor as our reference. Of course, this is always the contributor with all neutral atoms and alternating pi bonds. The name is constructed in much the same way as simple alkenes, but with the alteration before the ene suffix, we insert the prefix di, tri, tetra, etc., based upon the number of double bonds present.

For example, what's commonly called vinyl ethylene would be named 1, 3-butadiene. Now, creating ever larger linear conjugated alkenes produces 1, 3, 5-hexatriene, then 1, 3, 5, 7-octatetraene, and so on. Now, you may be wondering at this point if the analogy extends to cyclic conjugated compounds, and the answer to this question is yes. For example, 1, 3-cyclohexadiene would be considered a conjugated compound as well. Yet again, you may wonder if conjugation in cyclic compounds can run the full circumference of the ring, like in this putative 1, 3, 5-cyclohexatriene molecule. The answer to that question spawns such a rich chemistry that we're going to give it a lecture all its own next time. So for now, let's keep our focus on acyclic conjugated compounds and leave the more complex cyclics for another day.

Just as we did with isolated alkenes, we can also survey the effect of conjugation on stability. using heats of hydrogenation. So let's do this using a simple system with only a few isomers, like pentadienes. As a reference, the heat of hydrogenation of one pentene is 125 kilojoules per mole. Now, if we add a second double bond which is isolated from the other by multiple carbon-carbon bonds, we create 1, 4-pentadiene, which contains two isolated double bonds. Not surprisingly, the heat of hydrogenation doubles, because we've essentially added a second bond of exactly the same type. But, hydrogenation of its regioisomer, 1, 3-pentadiene, shown here in the trans conformation, releases less energy than the sum of that of two isolated double bonds. It's this observation which first led to our understanding of the stabilizing effect of resonance. This stabilizing effect extends far beyond simple diene hydrocarbons. Resonance has similar effects on the stability of compounds containing heteroatoms, like our familiar friends enols, carbocations, like allyl cations, and many others.

So let's take a closer look at our simple, conjugated diene from earlier and see if we can explain this remarkable increase in stability. We're going to begin by drawing it as what is clearly its most stable resonance contributor. Then we're going to turn on its side so that we can easily do the p-atomic orbitals, which are overlapping in space to produce the new conjugated pi system.

Now, if we try to create a representation of the resonance hybrid in this orientation, it becomes clear that something very special is happening in our conjugated system. The four pi electrons are not actually confined to those isolated pi bonds, but rather, have free run of the entire system. So, let's take a look at the molecular orbitals, first of a simple isolated alkene, and analyze their energetics so that we can better understand the stabilizing effect one more time.

I'm going to start by showing you a very simple p atomic orbital. And so far in the course, I've shown you these orbitals as simply having two lobes, one up, one down, one left, or one right, but always in opposition to one another. But the truth is that p-atomic orbitals have a characteristic called phase, and we tend to show this phase as being plus and minus. This is a rather unfortunate choice of terminology, because it really has nothing to do with charge or anything else that's up or down; it could just as easily be bitter or sweet, beautiful and ugly. They're just opposing phases, which means that I can draw my p orbital like this, or I can just as easily turn the phase the other way.

So all p orbitals have two potential states that they can be in. Their phases either up down, or down up, plus minus, or minus plus. Choose your language, but realize they have two opposing phases. Now, what this means is that if I put another p orbital next to it in an attempt to create a pi system, there actually are two different potential arrangements—an out-of-phase overlap. See? You have red blue, blue red. Or an in-phase overlap, where similar phases, red-red, blue-blue, are overlapping. Now, only orbital lobes that are in phase with one another can combine, so we get what's called a node in the out-of-phase overlap situation. This is a higher energy state, and so the orbital formed by this will be higher in energy. Whereas the in-phase overlap creates no node, and in fact, creates a larger space for the electrons to move, and this creates our pi bond.

Now, these are atomic orbitals. And the new orbitals that they form have shapes that look more like this, which traverse larger regions of space. But for today's discussion, to keep things simple, we're going to look more at the actual p-atomic orbitals before they've actually been combined, so we'll be using this representation, but we'll still be thinking about the large molecular orbitals as well.

So let's calculate now the energetics of the molecular orbitals in 1, 3-butadiene. So we know that our ethene had two different molecular orbitals. But what about 1, 3-butadiene? Well, its pi system has two pi bonds, and therefore, four electrons and four participating p orbitals. So I'm going to turn the molecule on its side here and place the p system on top. And then what we're going to do is we're going to analyze all the potential orientations, every permutation of phase up and phase down for these four orbitals combined. So let's start with the simplest. When all the orbitals are in phase, there are zero nodes, which means I've created a very stable region of space for the electrons to move it. So I'm going to put this at the bottom of my energy axis and label it pi 1. So this is the orientation for my lowest energy molecular orbital.

The next best arrangement has a single node right through the middle of the pi system. So this will be higher in energy than the previous orbital. We're going to call that pi 2. We can continue this exercise creating a new overlap that has two nodes. We're going to move that over and label it pi 3. And of course, the last potential permutation has three nodes, and we're going to move that to the highest energy position and call it pi 4. So, what you might notice here is that if I put all these orbitals on an energy axis, and I put the non-bonding energy here, which means, basically, the energy of a vacant p orbital, that some of the molecular orbitals are more stable, and others are less. So only when electrons move into these regions of space will they be happier than if they're not in a bond it all. To move into these orbitals would be energetically unfavorable.

Now fortunately, in the case of 1, 3-butadiene, we have four total pi electrons, which means they can all find a home in a lower energy orbital. This is the source of the stability of conjugated compounds; they offer a more stable, lower energy orbital for the electrons to move into when those p orbitals combine. Now, when they're down here in this region beneath, we call them bonding orbitals, for obvious reasons, because they create a chemical bond. But when we talk about orbitals up here above that non-bonding energy threshold, we refer to them as antibonding orbitals, because they're very high in energy.

But the effect is even more predictable and goes even further than that. We can look at it by comparing 1, 3-butadiene, which we just looked at, with a slightly larger conjugated alkene, 1, 3, 5-hexatriene. So if we do the same exercise for these two compounds that we just did, you will find that when you turn them on their side and analyze their p-atomic orbital overlaps, there are four molecular orbitals available for the butadiene, and in this case, there will actually be six different orbitals possible for the hexatriene.

But there's something important you have to notice here beyond just the number of orbitals. And that is that the distance between then in energy is changing. Here, the orbitals are closer together in energy than the others. So, if I take a look at these systems when populated, both of them create more stable compounds than they would be if they didn't have conjugation. So that's a good thing. But the other important effect that we can see here is that the frontier molecular orbitals, called the HOMO and LUMO, for highest occupied molecular orbital and lowest unoccupied molecular orbital, are different, not only different, predictably so. Extending conjugation decreases the frontier molecular orbital energy gap.

Now in the meantime, nature has exploited this property of conjugated compounds in the development of a rainbow of pigments used by plants to gather energy for photosynthesis, attract symbionic animals, and even to protect themselves from light, which could otherwise damage their cellular structure. Now, one of the most familiar examples of this from our everyday lives is the pigment beta carotene. Now, this is an example of the structure of beta carotene using ChemDraw. Just like our much smaller examples, beta carotene can absorb a very specific wavelength of light to promote electrons in its pi system from the HOMO to the LUMO. In this case, however, you can see that we have 22 p orbitals participating in its frontier molecular orbital energy gap. So that makes that gap so very small that it doesn't absorb ultraviolet light, but rather, low energy, visible light.

Now, if a specific color of light is absorbed when passing through or reflecting off of an object, even one as small as a molecule, our eye detects all the remaining wavelengths as a transmitted or reflected light, causing us to see the complementary color of that which is being absorbed. In this case, orange is the complementary color to the absorbed blue light. So that's the color that we see. Now, beta carotene, which is right here, is responsible in part for the color of fresh carrots, for which it was named, as well as a certain color changes in the leaves of trees in autumn. See, as fall comes, deciduous trees undergo a process called senescence, during which they break down and reabsorb the valuable chlorophyll which helps them to produce food through photosynthesis.

Now, as the chlorophyll is removed from the dim leaves, the other pigments, like this beta carotene, which were overwhelmed by its strong coloration, are given their chance to shine, even if only for a few weeks. As the leaf dies and detaches from the now-dormant tree. And it's these 22 participating p orbitals in the resonance of beta carotene that are responsible for much of the orange color that you see. So this year, when the fall foliage begins to burst with color, think about this. What you're really observing is the result of a complex dance of pi electrons within the pigments of the leaves.

Now, you might also recognize that beta carotene bears a striking similarity to another compound we've already looked at—retinol. And that's not a coincidence. Beta carotene is one of a group of compounds commonly referred to as Vitamin A, and when ingested, some animal species, including humans, can convert beta carotene into retinol.

But humans aren't very good at this. Our bodies make our own retinol in other ways as well. So we don't really have a need to convert large quantities of beta carotene into this critical chromophore, or molecule, which absorbs light. So what happens when a carrot lover just can't say no to that 20th or 30th carrot stick? Well, his or her body is unable to convert all the beta carotene into retinol, so instead, it stores the useful vitamin for a rainy day. Now, beta carotene, you probably noticed, is a hydrocarbon. So it's naturally stored in the lowest polarity environment that your body can find— fat tissue just beneath your skin. So if enough beta carotene is consumed, and therefore stored in this way, it can lead to a yellowing of the skin or even the appearance that the skin has turned orange. This condition, called carotenemia, is actually pretty harmless and will resolve itself over time with a minor dietary modification.

Although conjugated systems are somewhat more stable than similar but isolated pi systems, they can and do undergo many of the addition reactions that we discussed previously. For simple isolated alkenes, the same kind of reactions will apply to conjugated alkenes, but with a few important caveats. For example, the highly connected pi systems of conjugated compounds, like alkenes, means that they can often undergo more complex addition reactions with unexpected regioisomers forming, and the classic example of this is in the addition of a hydrogen halide to 1, 3-butadiene.

So let's take a look at the addition of this hydrogen halide to 1, 3-butadiene. We're all familiar with hydrochloric acid and its properties and how it tends to react with alkenes. So I'm going to put a 1, 3-butadiene here in place of where we've already seen some isolated alkenes in earlier lessons in the course. And your instinct is probably right here, to have your alkene attack the hydrogen of the HCL. But in this case, you get more than one potential intermediate. Notice that I could draw a carbocation with the positive charge on the secondary carbon here, or I could draw it on the primary carbon here. Now of course, the truth about this is that it's really all the same intermediate and that what we're really looking at in this slide is two different resonance contributors to what is actually a resonance hybrid with positive charges in two different locations. Nonetheless, this means an attack can occur by the chloride on one of two different spots on the intermediate. The chloride attacks the secondary carbon, we get the product that one would probably anticipate based upon our earlier lectures on isolated alkenes.

If instead it attacks the primary carbon, we expect a product in which the chlorine has added, actually, four atoms away instead of two atoms away from where we placed the hydrogen. So you can imagine in your mind that Zaitsev and Markovnikov would probably have a pretty good argument over this one, wouldn't they? Because Zaitsev would clearly say the more substituted alkene is what you're going to make. Whereas Markovnikov is more likely to argue that the more stable carbocation will dictate the product of your reaction. Now, although they didn't actually argue over this one, one can imagine it would be a pretty epic battle, wouldn't it? And the real catch here, the real clincher is, they would both be right; it just depends on the conditions. We call the first situation here a 1, 2 addition. This product tends to form very rapidly. We call the lower example a 1, 4 addition, and this product is actually more stable, but it also tends to form more slowly.

So we call these kinds of competitions between two different products which form under two different sets of conditions, at least in this case, a thermal dynamically or kinetically controlled reaction. So let me explain a little bit more. Here's our reaction again, 1, 3-butadiene with HCL. But this time, let's consider the vertical here as being an energy axis and track the reaction as it goes. So in the first step of my reaction, I'm going to form my intermediate, where I have positive charge in two different regions. Now, this is the same for either pathway, so I only have to draw one. But at this point, the pathway splits, and we go through two separate transition states, one with the chlorine attacking at the primary, and one with it attacking at the secondary. But notice that the intermediate which results in the attack on the secondary carbon is actually lower in energy.

If we complete the reaction, though, what we find is that the higher energy transition state has led us to a lower energy product. So the question here is, which is driving the major product of this reaction? Is it the stability of the transition states, meaning the rate of the reaction, or the stability of the products, meaning the thermodynamics, or the free energy of the reaction? So if we simplify this, it's a bit easier to see. We have a fast pathway to an unstable product and a slow pathway to a more stable product. So we call these kinetic products for that one which forms quickly, and a thermodynamic product for one which takes longer to form but is more stable.

So in the earlier decades of the 1900s, synthetic organic chemistry was earning recognition as a powerful field of study with tremendous potential for useful applications in medicine, industry, and beyond. But as the 20th century dawned, the library of synthetic techniques available to organic chemists was still somewhat lacking. A particular challenge facing chemists was that need for techniques to create new carbon-carbon bonds. Now, you may notice as you review the previous lectures that examples of techniques which can accomplish this critical task are scarce. In short, if you were a chemist in the early 1900s who could devise a new method for creating carbon-carbon bonds, you might as well start writing your Nobel Prize acceptance speech.

One proof of this proviso is the Diels-Alder reaction. Named for Otto Diels and Kurt Alder, who we met earlier. In 1928 they devised a reaction scheme which exploited the geometry of conjugated dienes and molecular orbitals to produce not just one new carbon-carbon bond, but two new carbon-carbon bonds in a single step.

The Diels-Alder reaction is actually one of a class of reactions known as pericyclic reactions. Pericyclic reactions consist of the exchange of multiple pi electrons simultaneously between two molecules to create two new sigma bonds between them. The Diels-Alder is designated a 4 plus 2 pericyclic reaction, because it involves this kind of a reaction between a conjugated diene, having four pi electrons, and one pi bond from another reagent known as dienophile, which would have 2 pi electrons.

The simplest example which meets these criteria is the reaction between the cisoid isomer of 1, 3-butadiene and ethene. So let's start there. To envision exactly how this reaction takes place, let's recreate the molecular orbitals of both molecules. Recall from our earlier discussion in this lecture, 1,3-butadiene has four molecular orbitals, of which the pi 2 is the highest occupied molecular orbital, or the HOMO. And the pi 3 is the lowest energy unoccupied molecular orbital, or the LUMO. Now, we collectively refer to the HOMO and LUMO as frontier molecular orbitals, because they present the orbitals which can be vacated and occupied respectively when a reaction take place. Ethene is, of course, a bit simpler since it only has two p orbitals in its pi system, so pi 1 is the HOMO, and pi 2 is the LUMO.

Now, this is the exciting part, where the elegant simplicity of this reaction shines through. As the two reagents approach one another, we can see that the HOMO of the diene and the LUMO of the dienophile are in phase with one another on both sides and that the same is actually true for the diene LUMO and the dienophile HOMO. Now, here we can see why the diene has to be cisoid as well. Transoid butadiene has the ends of its pi system too far part to achieve the needed overlap on both sides simultaneously. What this means is that the two reagents are perfectly configured to exchange electrons on both sides in a concerted process, and this produces two new sigma bonds from four pi electrons.

The entropic penalty of making one molecule from two is very easily balanced, and in fact, overwhelmed by the enthalpic benefit of trading those two pi bonds for the two sigmas. The new carbon-carbon bond creation is complete. What a rush, right? So what makes the 4 plus 2 reaction so easy to complete is this in-phase alignment of frontier molecular orbitals. Because the molecular orbital geometries are perfectly set up for a reaction to take place, all we need to do is mix the reagents and wait for the molecules to collide. So we call this a thermally activated reaction.

But if we try to conduct an analogous reaction, for example, in a 2 plus 2 cyclization, making cyclobutene from two molecules of ethylene, we see that the frontier molecular orbitals are not properly aligned when the compounds are in their lowest energy states. We're not completely out of luck, though; we can resort to a trick call photo activation promoting electrons into higher molecular orbitals by shining light of just the right wave length on them. This essentially changes the identity of the frontier molecular orbitals, artificially creating the overlap that we need to get a 4 plus 2 addition. Now, it's this detail that the Diels-Alder reactions are thermally activated that makes them so useful and also the reason that I'm going to stick to the 4 plus 2 reaction as we go forward.

Now, it also doesn't hurt that a number of biologically and commercially relevant compounds, such as cyclic terpenes, contain a cyclohexene motif. Compounds naturally occurring in such plants as the limonene in oranges, the terpenol in lilac, and the alpha ionone from violet can all be easily synthesized using Diels-Alder strategies.

Of course, the Diels-Alder reaction can be performed with more complex dienes and dienophiles, leading to products with more robust stereochemistry than our cyclohexene product from the previous example. Notice that in the addition of a dienophile to the diene, both of the new carbon-carbon bonds form on the same side of the newly created ring. In the case of cyclohexene formation, this is purely an academic observation, since all substituents of the newly joined carbons are hydrogen atoms. But what if they're not?

To find a situation in which this applies, we need to look no further than the very first step of the Woodward synthesis of cholesterol. Woodward knew that he was trying to make a compound with four fused carbon rings on it from much simpler starting materials. So why not consider the Diels-Alder to start building a scaffold? Why not indeed. Woodward began his synthesis using 1, 3-butadiene to create his first fused ring to a compound known as hydroquinone. The Diels-Alder process ensured that the desired six-membered ring would form. But Woodward had a new problem in his synthesis; the newly fused rings had the proper connectivity but were in the cis conformation, and clearly, his final product had to be in the trans conformation. We'll see how he dealt with this issue a little bit later.

Today though, we've covered a great deal, so let's sum up. We discussed the phenomenon of conjugation in which multiple pi bonds in resonance with one another lend extra stability to a compound or ion. We made a few structural arguments for why this added stability exists, and we learned a new way of modeling the energies of conjugated pi systems in alkenes, backing up our predictions of stability with observed heats of hydration

We scratched the surface of conjugated diene reactivity by exploring two reactions, first, the hydrohalogenation of 1, 3-butadiene, and how addition of the second double bond leads to two possible products—a kinetic one, the 1, 2 product, and the thermodynamic one in the 1, 4 product. Then we examined a Nobel-Prize-winning reaction conceived by Otto Diels and Kurt Alder. A reaction which allows us to quickly and easily form not one but two carbon-carbon bonds.

We discussed how the frontier molecular orbitals of reagents in the 4 plus 2 pericyclic reaction are perfectly suited for a thermally activated reaction, which can easily produce a motif which got Woodward started on his way to the total synthesis of cholesterol.

Next time, we're going to take a look at what happens when we take these linear conjugated systems to the next level by cyclizing them to produce a very special carbon scaffold. I'll see you then.

Benzene and Aromatic Compounds
Lecture 20

This lecture focuses on characterizing and naming a class of compounds known as aromatics. You will learn how cyclic pi systems of $4n + 2$ pi electrons have a remarkable stability that was recognized very early in the history of organic chemistry. In addition, you will learn about Hückel's rule, which predicts that this trend continues on to larger systems, and the polygon rule. You also will examine how aromaticity can profoundly influence the acid–base properties of organics. Finally, you will consider situations in which aromatic rings fuse together to create a new class called polynuclear aromatic hydrocarbons.

The Structure of Benzene

- It was known for quite some time that a compound of unusual stability and a very low hydrogen-to-carbon ratio could be isolated from torch fuel. Although the compound was clearly a hydrocarbon, its unusual chemical stability and reduced hydrogen content defied explanation until the 1860s, when August Kekule hypothesized that benzene's formula and properties could be explained if the molecule were cyclic.

- We know today that Kekule was right. His theory was not beyond challenge, however, as Kekule's model of a cyclic triene predicts that a molecule of benzene should have two different types of carbon-carbon bonds: three shorter double bonds and three longer single bonds. Yet the compound proved to have six identical bonds, forming a perfect hexagon.

- Also, Kekule's putative structure should generate two possible isomers when replacing two of the hydrogens with other groups, such as chlorine. One isomer has a double-bonded carbon between the substituents, and the other has only a single bond. Kekule attempted to explain this observation as a fast interconversion between the two putative isomers, but he was actually wrong about this.

- In 1925, Sir Robert Robinson published a theory in the *Journal of the Chemical Society* postulating that what he called an "aromatic sextet" of electrons stabilized the ring by moving unrestricted about the entire circular pi system, creating a ring of identical carbons joined by bonds of one and one-half order. This theory leads us to more accurately draw benzene molecules using a circle or dashed interior hexagon to represent the aromatic sextet of electrons.

- In the 21st century, far from just qualitatively observing the stability of benzene, we can use the technique of calorimetry to determine its heat of hydrogenation—this is, the amount of heat released when the unsaturated compound is saturated with hydrogen. If we conduct this experiment using cyclohexene and 1,3-cyclohexadiene as controls, we can determine with relative ease that the double bonds in these systems are nearly equivalent.

- But when we add that magical third pair of pi electrons, the heat of hydrogenation is far less than we would have expected—about 150 kilojoules per mole less. We call this discrepancy the resonance energy for benzene. To put the size of that resonance energy into perspective, it is just less than half the bond enthalpy of a covalent carbon-carbon bond.

Molecular Orbitals of Benzene, Hückel's Rule, and Frost Circles
- Benzene consists of six sp^2 carbon atoms joined in a ring. This ring is perfectly flat in the case of benzene, which allows all of its p atomic orbitals to combine to form one large pi system with density both above and below the ring.

- But the rule for the formation of molecular orbitals is that six atomic orbitals in means six molecular orbitals out. These orbitals all correspond to the linear addition of p atomic orbitals with varying combinations of phase. For example, when all six p orbitals are in phase, the lowest possible energy molecular orbital is obtained. We call this the pi 1.

- But it is also possible to form five other permutations of molecular orbitals: two possibilities with a single nodal plane (pi 2 and pi 3), two possibilities with two node planes (pi 4 and pi 5), and one possibility in which all six p orbitals are aligned antiparallel, creating three nodal planes and a sixth molecular orbital (pi 6).

- So, we are in a situation very similar to that of 1,3,5-hexatriene, with the caveat that some of the molecular orbitals are equivalent in energy. If we orient these molecular orbitals on a vertical energy axis, we can superimpose a six-membered polygon on the energy diagram for a six-membered ring.

- Bisecting this ring with a horizontal line gives us the nonbonding energy for the diagram, and populating the pi system with the total inventory of six pi electrons from benzene leads us to a situation in which all of those electrons are in a more stable bonding molecular orbital. This is the source of benzene's unusual stability.

- In fact, this geometric coincidence—called the polygon rule or a Frost circle—holds for all simple annulene structures like this one. In fact, they can be used to explain yet another concept, known as anti-aromaticity.

- The molecular orbital diagram for cyclobutadiene predicts that moving its four pi electrons into a planar system requires that two of them populate nonbonding orbitals. This arrangement leads not to an energetic benefit, but rather to a penalty for aligning all four p orbitals parallel to one another.

- It is actually more thermodynamically favorable for cyclobutadiene to have two isolated double bonds. Without any aromaticity to offset the tremendous angle and torsional strain it is under, cyclobutadiene is a hopeless proposition.

- Consider the next progression in this type of molecule: 1,3,5,7-cyclooctatetraene. If we inscribe an octagon within a circle, point down, there will be eight pi molecular orbitals in the system for this compound. When we populate these orbitals with eight electrons from its pi system, we see a situation similar to that of cyclobutadiene, in which the last two electrons are in nonbonding orbitals, making 1,3,5,7-cyclooctatetraene anti-aromatic.

- This fact is evident in the structure of 1,3,5,7-cyclooctatetraene, which has been shown to have four isolated double bonds of shorter length and a distinct pucker, which places alternating pi bonds out of alignment with one another in an attempt to avoid anti-aromaticity.

- If we continue our analysis by expanding the pi system yet again, we see that 1,3,5,7,9-cyclodecapentene is again aromatic, and so on. This trend—that $4n$ pi electrons leads to anti-aromatic systems and $4n$ pi + 2 electrons leads to aromatic systems—is known as Hückel's rule, named for German chemist Erich Hückel.

- But the utility of Hückel's rule extends well beyond just hydrocarbons like benzene; it can be used to explain the behavior of other cyclic systems with aligned p orbitals as well. For example, a polygon analysis of the tropylium ion, which is a carbocation consisting of a seven-membered ring with six pi electrons and a positive charge, shows that it is also aromatic.

- In fact, the effect of resonance on the stability of tropylium is so powerful that it is one of the few carbocations that can be purchased as a stable salt and stored in the laboratory. Those pesky intermediates from first-order reactions have become a starting material, owing to the stabilization provided by resonance. And the same is true for cyclopentadienyl cation, which contains six pi electrons and a system of five pi molecular orbitals. In both instances, all six pi electrons find homes in bonding orbitals, lending extra stability to the ions.

- Hückel's rule can be extended to systems containing heteroatoms— for example, pyridine (C_4H_5N).

Naming Aromatic Compounds
- The revelation that all carbons of benzene are equivalent simplifies the process of naming them when they are substituted. Because all six carbons are identical and the pi-bonding electron density is the same, using the name 1,3,5-cyclohexatriene (per IUPAC) is a bit misleading, so we usually default to describing this ring motif as "benzene."

- Just as any other carbon scaffold can be modified by replacing its hydrogen atoms with more complex substituents and functional groups, so can those of benzene.

- When a single substituent is attached to a benzene ring, we simply start the name with that substituent's prefix—for example, chlorobenzene, hydroxybenzene, or methylbenzene. Because all six carbons of benzene are equivalent, the position of the first modification is of no consequence. We get "chlorobenzene" no matter which carbon we attach it to.

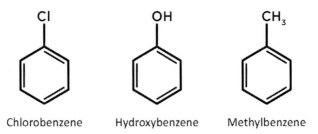

Chlorobenzene   Hydroxybenzene   Methylbenzene

- Attaching a single functional group to the ring prompts us to place a modifier at the end of the name—for example, benzaldehyde or benzoic acid.

- Sometimes the groups and the parent get turned on their head, and we describe the benzene ring as a substituent itself, using the identifier "phenyl" or "aryl." For example, chrlorobenzene is also often called "aryl chloride," and hydroxybenzene is called "phenol."

29

- But what really spices up the nomenclature of benzene rings is when we have two or more attached substituents and functional groups, because the attachment of a single substituent renders the remaining unmodified carbons different from one another.

- Take methylbenzene, for example. There are three chemically distinct locations on the ring. One of them is a single bond away, another is two bonds away, and a final position is three bonds away.

- Where we place the second group makes a profound difference in the identity and chemistry of the compound. For example, dimethylbenzene, commonly called xylene, can form three different isomers: 1,2-dimethylbenzene, 1,3-dimethylbenzene, and 1,4-dimethylbenzene.

- We often use a more common method of nomenclature to indicate which regioisomer of the compound we are discussing, using the prefixes "ortho-," "meta-," and "para-" in place of the numbered locations of the groups. The prefixes are often shortened to a single italicized letter—for example, the three isomers o-xylene, m-xylene, and p-xylene.

Large Polynuclear Aromatics

- Aromatics need not exist as an isolated benzene ring. Although most, if not all, of the compounds we will be investigating will have isolated benzene motifs, a newer class of compounds has gathered considerable interest in recent decades.

- Picture two benzene rings fused along one edge, essentially sharing a carbon-carbon bond. We call this compound naphthalene. By fusing the rings, we create two aromatic systems that share a common pair of p orbitals, so it should come as no surprise that naphthalene's resonance energy is about 252 kilojoules per mole—slightly less than two separate benzene rings, but substantial nonetheless.

- For planar compounds, this trend continues through the tricyclics anthracene and phenanthrene all the way up to graphite, which is essentially a never-ending sheet of fully fused benzene rings. This allotrope of carbon benefits from such significant resonance stabilization that at standard temperature and pressure, it is even more stable than diamond.

- In the 1980s, chemists began devising ways to endow curvature to these so-called graphene sheets by changing which edges each ring shares with others, creating new allotropes of carbon in the form of spheres. The most famous of these is the truncated icosahedron C_{60} molecule buckminsterfullerene, affectionately referred to as "buckyballs" and named for the famous architect, inventor, and futurist R. Buckminster Fuller.

- One can also envision how the buckyball motif could act as endcaps connecting a graphene sheet that is rolled up on itself. This structure is commonly called a carbon nanotube. These materials show some promise in some very exciting applications—such as gas storage, media, and building materials to electronics, engineering, and medicine—but much of that potential is yet to be realized.

Suggested Reading

Berson, *Chemical Creativity*, Chap. 3.

Davenport, D. A., "Observations on Faraday as Organic Chemist Manqué," *Bulletin for the History of Chemistry* 11 (1991): 60–65.

McMurry, *Fundamentals of Organic Chemistry*, Chaps. 5.1, 5.2.

Thomas, J. M., "The Royal Institution & Michael Faraday: A Personal View," *Bulletin for the History of Chemistry* 11 (1990): 4–9.

Wade, *Organic Chemistry*, Chaps. 16.1–16.14.

1. Which of the species depicted below are aromatic?

2. Imidazole is a frequently used organic base. Which nitrogen from imidazole is more basic, and why?

Benzene and Aromatic Compounds
Lecture 20—Transcript

You may recall from our discussion on alkenes, that the discovery of petroleum took place around 1860, drastically altering the methods used to recover and refine fuel oils. To begin today's discussion, we'll have to move farther back in time, several decades before Drake's discovery of petroleum, to a time when fuel oils were still sourced from whale and fish oil.

Now, Michael Faraday was an English researcher; he's most well known for his work in the field of electromagnetism. But in 1820, he developed a keen interest in the chemistry of fuels. So interested was Michael Faraday, that he all but abandoned his very successful line of research in electromagnetism so that he could begin deconstructing the composition of, what was then commonly known as illuminating gas, a processed form of fish or whale oil commonly used in the lamps of the day.

Faraday worked for nearly five years refining his techniques for isolation of certain components from that material. Most notable among them was a liquid, which he found he could produce by pressurizing the gas to 30 atmospheres and isolating the condensate which formed. Even in Faraday's time, it was known that hydrocarbons with suboptimal amounts of hydrogen were susceptible to reactions with such reagents as chlorine gas or molecular bromine. So, Faraday dutifully determined the hydrogen-to-carbon ratio in his isolate, finding that it was extremely low.

Now in his day, scientists thought that carbon had an atomic mass only half that of what it actually is, so Faraday erroneously named his new discovery bicarburate of hydrogen. In fact, carburate hydrogen, or monocarburate of hydrogen, would have been more appropriate, but considering the circumstances, I think we can forgive this little miscalculation. And, this error not withstanding, when comparing Faraday's bicarburate of hydrogen formula to other hydrocarbons of the day, it was very clear that his new compound had a very low hydrogen-to-carbon ratio.

So, Farady had already isolated his new compound, bicarburate of hydrogen, from his lamp oil, but he wanted to go about doing some more analysis.

Now, he had combusted this material already, measuring how much carbon dioxide and water formed in the process, which gave him a rough idea of the formulas; now, he was off by a factor of two because of the putative mass of carbon in that day, but I've corrected that in my formulas here for my demonstration.

So, what I have here, as soon as I protect myself properly from it, are some vials containing molecular bromine. These colorful solutions here are Br_2, contained in an inert solvent, methylene chloride. I also have samples of some of the hydrocarbons that had been isolated in Faraday's time, and those include C_6H_{14}, C_6H_{12}, C_6H_{10}, C_6H_8, and C_6H_6. Again, I've corrected these formulas for the actual mass of carbon.

Now, I'm going to cover my hands with something a little bit to make sure that they're safe as well. And I'm going to recreate the experiment that Farady would have done in his day, which generated such a buzz about his new bicarburate of hydrogen. What I'm going to do here first is I'm going to uncap my bromine solutions, and I'm going to add them to the various hydrocarbons that I have placed here as well. And when I do that, if a reaction takes place between the bromine and the hydrocarbon, naturally, we would expect to the bromine to be something other than bromine, and the color will change.

So, let's start with the first hydrocarbon C_6H_{14}. So I add my bromine. Be sure that we're well agitated. You may notice that no reaction's taken place. Now, this was well known to Faraday. Now, he also knew that if you took this other compound here, C_6H_{12}, which had a slightly lower hydrogen-to-carbon ratio, and add it, again, molecular bromine, that a reaction took place. Clearly, no pun intended, a reaction has happened here. Move my bromine vial to the back.

Now, he could continue this experiment going along all the various isolates from all the various sources available to him, including C_6H_{10}. We have our reaction, gone clear. One more. Let's uncap our C_6H_8. Again, an unsaturated hydrocarbon, and a reaction takes place.

Now, Faraday knew all of this was happening. All of these experiments had been done previously. So when he isolated a new compound to determine that it's formula had even less hydrogen, he clearly predicted, and everyone thought he'd be right, that this new bicarburate of hydrogen would also react with molecular bromine. So, just like any good scientist would, he validated that by adding molecular bromine. Give it a swirl to be sure that we're well mixed. We have a problem, don't we? The trend is bucked here. The C_6H_6 doesn't react the same way as all of the other hydrocarbons with suboptimal hydrogen to carbon ratios.

So, Faraday found this fascinating, and he needed to figure out what was going on. Alas, Faraday never completed his characterization of this new compound. And sadly for us chemists, Faraday's first love of electromagnetism took hold again and distracted him from his curious hydrocarbon. So, Faraday had to settle for inventing electromagnetic induction, which essentially heralded the arrival of the age of widely useful electricity. Not a bad consolation prize, I suppose.

Instead the world would have to wait until the 1860s when August Kekule, a German-born scientist working in Belgium, took up the mantle of explaining the low hydrogen content and unusual stability of this compound, which had since been renamed benzene. Now, this guy's a real looker, isn't he? This is how I know I would never make it as an Eastern European chemist in the 1800s. I can't pull off a beard that good.

So, it's known for quite some time, that a compound of unusual stability and a very low hydrogen-to-carbon ratio could be isolated from torch fuel. Although the compound was clearly a hydrocarbon, it's unusual chemical stability and reduced hydrogen content defied explanation. At least until Kekule had, what he claimed in later years, was a genuine revelatory dream. In an 1890 interview, near the end of his career, Kekule claimed that he had this famous vision while dosing on the upper deck of a horse-drawn bus in 1850s London. As he tells it, he dreamt of a snake biting its own tail and awoke to the revelation of benzene's formula and properties could be explained if the molecule were, in fact, cyclic.

There are many who wonder to this day why Kekule would wait nearly four decades to tell such a fantastic tale. In fact, most doubt the story's validity, but the story is so colorful and inspirational that it lives on in texts to this day. So, whether it came by divine revelation or educated guess, we may never know. But we do know today, Kekule was right. What Kekule had proposed was this, that we have a molecule, which is cyclic, containing spaced double bonds around that molecule, and this would explain why it would have different chemistry than all of the other conjugated alkenes, or hydrocarbons, as they would call them in his day.

Now, when you take a look at the structure, I want you to notice something important. Kekule's theory about cyclization of a hexatriene is not beyond reproach, because benzene itself would have three double bonds if they were isolated, leading to the potential for two different isomers. See, once scientists were able to figure out how to attach halogens, there are actually two potential arrangements for the addition of a halogen to a benzene ring, or more accurately, a substitution. We can put two halogens at the top here, where the double bond is located, meaning that our halogens would be in between a double bond. But, if I wanted to, I could place those halogens on opposite sides of a single bond instead. So, if Kekule's theory was right, that this was simply a cyclic hexatriene, then there should be two isolatable isomers of, say, dichlorcyclohexatrine, and yet, there were not. An interesting conundrum, isn't it? Now, Kekule attempted to explain this observation as a fast interconversion between the two putative isomers. But there was no revelatory dream to save him this time; he was actually wrong about that.

The story of benzene was completed in 1925, when Sir Robert Robinson published a paper in the Journal of the Chemical Society postulating that what he called an aromatic sextet of electrons stabilized the ring by moving unrestricted about the entire circular pi system. This would create a ring of identical carbons joined by bonds of 1½ order and would solve the problem of why there aren't two different isomers. This theory leads us to more accurately draw benzene molecules using a circle or dashed interior hexagon to represent the aromatic sexton electrons. But it isn't the early 1900s; it's the 21st century, and we have much more sophisticated techniques than Faraday, Kekule, and even Robinson had. So let's bring those to bear on the issue of benzene stability.

Far from just qualitatively observing the stability of benzene, we can use our familiar technique of calorimetry to determine its heat of hydrogenation. This is the amount of heat released when the unsaturated compound is completely saturated with hydrogen. If we conduct this experiment using cyclohexine, 1,3-cyclohexadiene also as controls, we can determine with relative ease, that the double bonds in these systems are nearly equivalent to one another. But, when we add that magical third para pi electrons, the heat of hydrogenation is far less than we would have expected, about 150 kilojoules per mole less. We call this discrepancy the resonance energy for benzene. And to put the size of that resonance energy into perspective, it's just less than half the bond enthalpy of a covalent carbon-carbon bond.

So about a century passed between Faraday's first isolation and characterization of benzene and Robinson's final description of the aromaticity that it possesses. And this opened the floodgates, and it wasn't long after this that the molecular orbital energetics of benzene and other of cyclic molecules with alternating double bonds, which we collectively call annulenes, were worked out by a German chemist, named Eric Huckel. You see, just as we produced mathematical combinations of p-atomic orbitals to model pi systems in linear conjugated compounds last time, we can use the same technique for cyclic alkenes with overlapping pi bonds.

Let's start by sticking with the familiar aromatic compound, benzene. Benzene consists of six sp^2 carbon atoms joined in a ring. Now, this ring is perfectly flat in the case of benzene, which allows all of it's p-atomic orbitals to combine to form one large pi system with density both above and below the ring. But remember our rules for the formation of molecular orbitals— six atomic orbitals in means six molecular orbitals out. These orbitals all correspond to the linear addition of p-atomic orbitals with varying combinations of phase.

For example, when all six p orbitals are in phase, the lowest possible energy molecular orbital is obtained. We call this the pi 1. But it's also possible to form five other permutations of molecular orbitals. Two possibilities exist with a single nodal plane; let's call them the pi 2 and the pi 3. Two possibilities with two nodes planes also exist, so let's call those the pi 4 and pi 5. Finally, there's one possibility in which all six p orbitals are aligned anti parallel, creating three nodal planes and a sixth possible molecular orbital, pi 6.

So let's take a look now at aromaticity and our famous popular benzene molecule as well as a different cyclic compound with alternating pi bonds. And what we're going to do now is analyze the energetics of those pi molecular orbitals that we just looked at, not only for benezene, but also for a different cyclic compound with multiple pi bonds. Let's use 1,3-cyclobutadiene and benzene as our subjects. So here, I've drawn 1,3-cyclobutadiene with two isolated pi bonds. And yet I've drawn benzene with three delocalized pi bonds, denoted by the dashes moving all the way around the ring. And that's not an accident.

So, if we think about the energetics of those six molecular orbitals, from benzene to start, there's a little mnemonic trick we can use to determine the relative energies of those orbitals. What we do is we draw a circle on an energy axis, and we inscribe within that circle, point down, a polygon with a number of sides equal to that the cyclic compound we'd like to analyze. For example, benzene having six atoms in the ring is going to have six corners it it's polygon, with one corner pointed straight down.

When I do this, the corners of the polygon where they touch the circle, indicate the relative energies of the six pi-molecular orbitals of this particular compound. So if I draw my non-bonding energy line through the center and populate it, what I discover is that the bonding region gets all of the electrons from the pi system, all six of them. So clearly, benzene's perspective on this molecular orbital system is that it would like to have it. I'd like to populate this. This is great because all my electrons get to go down in energy.

But what if we do a similar analysis for 1,3-cyclobutadiene, which is another annulene which looks, for all intents and purposes, like it should be aromatic. Well, if we do the exact, same exercise, inscribing our polygon and forming our molecular orbital energies, we find something quite different. We find that when we populate it with the four pi electrons from cyclobutadiene, the two of them go into non-bonding molecular orbitals. In other words, there's little, or in this case, really, no energetic benefit to this particular molecule being planar and forming these new molecular orbitals.

But the trend actually continues; it's a predictable trend in the stability of annulenes. So we've already seen 1,3-cyclobutadiene in benzene treated in this way. So let's go on through those, create their polygons, and populate them. Benzene, or cyclobutadiene, rather, we know is anti-aromatic. If we do the same exercise for benzene, as we did previously, we see that benzene is aromatic because of the extra stability. Cyclooctatetraene, on the other hand, will actually populate non-bonding orbitals if it's planar. Whereas, cyclodecapentaene will not, making it aromatic.

So you're probably beginning to notice a trend here. Well, you're not the first. Huckel was the first; that's why he was the one who got to name it. Sorry. But, what it turns out is happening here is, that every time we add 2 pi electrons, we switch from anti aromatic to aromatic to anti aromatic to aromatic, and therefore, we can model this as a simple mathematical equation, 4 n pi electrons leads to an anti-aromatic system, while 4 n plus 2 electrons leads to an aromatic system. This is the origin of Huckel's Rule.

But the utility of Huckel's Rule extends well beyond just hydrocarbons like benzene. It can be used to explain the behavior of other cyclic systems with aligned p orbitals as well. For example, the tropylium ion, which is a carbocation consisting of a seven-membered ring with six pi electrons and a positive charge. Polygon analysis of this species shows us that it's also aromatic. In fact, the effect of residence on the stability of tropylium is so powerful that it's one of the few carbocations which can be purchased as a stable salt and stored on a shelf in the laboratory. That's right, those pesky intermediates from first-order reactions, have become a starting material, owing to the stabilization provided by this very special type of resonance. So, cyclopentadienialanion also contains six pi electrons, and this time, a system of five pi molecular orbitals. And you can clearly see that in both instances all six pi electrons find homes in bonding orbitals, lending extra stability to the ions.

But we don't need to stop at hydrocarbons and their ions. Huckel's rule can be extended even farther to systems containing heteroatoms, as well. Take the example of pyritine, C_4H_5N. Now, when I draw this molecule in class, I usually have more than a few hands go up, and that's because as I've drawn it, the lone paranitrogen seemed to be in an sp^3 orbital isolated on the nitrogen atom. And you could argue this molecule can't be aromatic as drawn, because the pi system isn't cyclic. And you'd be right.

To properly depict pyridine, I have to move that lone pair into a p atomic orbital, allowing it to join that pi system of the ring. And in doing so, I produce an aromatic system in which the amine nitrogen and hydrogen are both coplanar with the rest of the molecule. The energetic benefit of aromaticity exceeds that of orbital hybridization in this situation. The effects of aromaticity on the basicity of pyrrole are also dramatic. Note that accepting a proton for pyrrole mean sacrificing that stabilizing aromiticity. Well, the same process for it's close chemical cousin, pyridine, does not. So it should come as no surprise that pyrrole is literally millions of times less basic than pyridine, all a powerful testament to the influence of aromaticity, and also, a very obvious application of Huckel's rule in action.

There are many other examples of aeromatics which contain heteroatoms, like the mauvine that we encountered in our nitrogen lecture. But for today, we're going to turn our attention back to hydrocarbons to keep our discussion focused. Now, just as with all other molecules, if we want to have meaningful discussions about benzene and its derivatives, then we need to agree on a nomenclature standard. The revelation that all carbons of benzene are equivalent simplifies the process of naming them when they're substituted. Since all six carbons are identical and the pi bonding electron density is the same, using the name 1,3,5-cyclohexatryene per the IUPAC's standard, is a bit misleading. So, we usually default to describing this ring motif as benzene.

Now, just as any other hydrocarbon scaffold can be modified by replacing its hydrogen atoms with more complex substituants and functional groups, so can benzene. We'll take a look at some techniques used to accomplish these modifications next time, but for now, let's focus on the nomenclature of the modified compounds.

When a single substituent is attached to a benzene ring, we simply start the name with that substituent's prefix. For example, chlorobenzene, hydroxybenzene, or methylbenzene. Since all six carbons of benzene are equivalent, the position of that first modification is of no consequence. We get chlorobenzene no matter which carbon I attach my chlorine to. Attaching a single functional group to the ring prompts us to place a modifier at the end of the name. Well, for example, benzaldehyde, or benzoic acid. Now, if this distinction between substituents and functional groups isn't confusing enough, sometimes the groups and the parent get turned on their head, and we describe the benzene ring as a substituent. So we would use the identifier phenal or aryl in situations like this. For example, chlorobenzene is also often called aryl chloride, and hydroxybenzene sometimes called phenol.

But what really spices up the nomenclature of benzene rings is when we have two or more attached substituents in functional groups, since the attachment of a single substituent renders the remaining unmodified carbons different from one another. Let's take a look at methyl benzene. There are now three chemically distinct locations on the ring. One of them is just a bond away; another, two bonds away; and the final position, three bonds away on the opposite side. So, where we place the second group makes a profound difference in the identity and chemistry of the new compound.

For example, dimethylbenzene, commonly called xylene, can form three different isomers, 1,2-dimethylbenzene; 1,3-dimethylbenzene; and 1,4-dimethylbenzene. Now, often we use a more common method of nomenclature to indicate which regioisomer of the compound we're discussing, using prefixes ortho, meta, and para in place of the numbered locations on the groups. The prefixes are often shortened to a single italicized letter. So here we would describe these three isomers as o-xylene, m-xylene, and p-xylene.

This nomenclature system is the source of the para in the popular pain reliever, paracetamol, also known as acetaminophen. Now, notice that the two groups in this compound are attached to the benzene ring in the para arrangement; it's only when they're at that specific position relative to one another, that the drug works. In fact, during the industrial-scale synthesis of acetaminophen, a portion of the phenol starting material is converted to ortho nitrophenol, which would lead to the ortho isomer of acetaminophen in a completed synthesis. But this orthonitrophenol is removed before continuing, because it leads to an inactive isomer of the compound.

It's worth mentioning that aromatics need not exist as an isolated benzene ring. Though most, if not all, of the compounds we'll be investigating will have isolated benzene motifs, a newer class of compounds has gathered considerable interest in recent decades. Picture two benzene rings fused along one edge, essentially, sharing a carbon-carbon bond. Now, this is not a new compound here; we call this compound naphthalene, and we've known about it for quite some time. Notice that by fusing the rings, we've created two aromatic systems which share a common para p orbitals. So it should come as no surprise that naphtalene's resonance energy is about 252 kilojoules per mole, slightly less than two separate benzene rings, but substantial, nonetheless. For planar compounds, this trend continues, through tricyclics, anthracene, and phenanthrene, all the way up to graphite, which is essentially a never-ending sheet a fully fused benzene rings. Now this allotrope of carbon benefits from such a significant resonance stabilization that at standard temperatures and pressures, it's even more stable than diamond.

In the 1980s, chemists began devising ways to endow curvature to these so-called graphene sheets by changing which edges each ring shares with others. This would create new allotropes of carbon in the form of spheres. The most famous of these is, hands down, the truncated iccosahedron C60 molecule, known as buckminsterfullerene, or buckyballs. These were named for a famous architect, and inventor, and futurist—Buckminster Fuller—who was very fond of this geometric arrangement.

One can also envision how the buckyball motif might act as an end cap, connecting a graphene sheet, which is rolled up on itself. Now this the structure is commonly called a carbon nanotube. And these materials show some promise in some very exciting applications such as gas storage media in building materials, electronics, engineering, and in medicine as well. But much of the potential is yet to be realized. Even if we never find a use for these, though, I think we can all agree they are just beautiful to look at, aren't they?

And yet another useful material created from graphene sheets is what's commonly known as carbon fiber. Most of us are probably familiar with this material, which was first created in the '80s, and has since found its way into a host of applications. Many of these applications exploit carbon fiber's enormous tensile strength, which rivals that of steel, but in a material which is four times less dense. As opposed to graphite, which consists of large, planar graphene sheets stacked in a regular, repeating structure, carbon fiber can be thought of as a series of graphene sheets which are bunched and tangled together with one another. This bunching and tangling prevent the graphene sheets from sliding along one another, creating a material with incredible tensile strength, rather than soft, lubricating properties, like graphite.

So today we focused on characterizing in naming a class of compounds known as aromatics. we discussed how cyclic pi systems of 4n+2 pi electrons have a remarkable stability, which was recognized very early in the history of organic chemistry.

We covered Huckel's Rule, which predicts that this trend continues on to larger systems as well. We examined what's commonly called the polygon rule, an extension of a mnemonic we use for linear conjugated alkenes, which is used to construct energy diagrams for cyclic pi systems. We also used this tool to demonstrate the principles on which Huckel's Rule is founded. We took a look at how aromaticity can profoundly influence the acid base properties of organics by stabilizing conjugate acids and bases. Particularly, we looked at the example of pyrrolidine in comparison to pyridine.

Next, we tried our hand at naming some of these compounds, placing substituents ahead of benzene and functional groups after benzene in the names, and how we use the prefixes ortho, meta, and para to describe di-substituted benzene rigioisomers.

We then concluded by considering situations in which aromatic rings fused together to create a new class called polynuclear aromatic hydrocarbons, and how this naturally extends to buckminsterfullerenes, and carbon nanotubes, and carbon fibers, which are the focus of current research in many labs around the world.

Next time we'll take a look at some of the techniques that organic chemists use to create these modified benzene rings. And I'll see you then.

Modifying Benzene—Aromatic Substitution
Lecture 21

This lecture will build on your understanding of the structure of benzene by investigating a very useful class of reactions known as electrophilic aromatic substitution reactions. You will investigate the general mechanism by which it takes place, and you will learn how other reagents often need to be activated to make them reactive enough to entice the substantially stabilized benzene ring to react with them. In addition, you will learn about a handful of the many modifications that can be made to benzene. Finally, you will consider how already-modified benzenes can be further modified to produce ortho, meta, and para isomers.

Electrophilic Aromatic Substitution

- In 1873, at the University of Strasbourg in Germany, Adolf Von Baeyer was supervising a graduate student named Othmar Zeidler, whose graduate project included the synthesis of a substituted benzene compound known as dichlorodiphenyltrichloroethane (DDT). Zeidler's synthesis of this compound was accomplished using chlorobenzene and chloral under acidic conditions. Under these conditions, the two chlorobenzene rings are joined by substituting the chloral for hydrogens in the para position of each ring.

- The Swiss chemist Paul Muller is credited with later discovering, documenting, and obtaining a Swiss patent for the insecticidal use of the compound, after which the U.S. government tested and used the compound to combat infectious disease abroad. Shortly after the conclusion of World War II, the U.S. government began to allow commercial use of DDT for pest control in domestic crop production.

- Unfortunately, DDT lasted so long in the environment that it could be transmitted from insects to fish, and from fish to birds. Many bird species suffered tremendous population declines due to the presence of DDT in their food sources. In 1972, the United States banned the use of DDT in agricultural applications. Today, it only finds use in emergencies as a vector control agent to prevent the spread of disease.

- The synthesis of DDT is a shining example of an electrophilic aromatic substitution reaction, which is so named because the newly attached motif starts out as an electrophile (unlike nucleophilic substitution, in which the newly added group is nucleophilic).

- In the reaction of benzene and a generic electrophile, for example, the pi electrons of benzene are not held as tightly as most sigma-bonding electrons, even though they are in an aromatic system. Although it takes an extraordinarily electrophilic reagent to coax them into attacking, it can be accomplished.

Though now banned in the United States, DDT is a shining example of an electrophilic aromatic substitution reaction.

- When an adequately electrophilic reagent is added to benzene, the pi electrons of benzene simply can't resist attacking, and in doing so, they generate a tetrahedral center adjacent to a resonance-stabilized carbocation. We call this intermediate a sigma complex because we have traded two of our aromatic pi electrons to make a new sigma bond to the electrophile.

- Even with the newly formed, more stable sigma bond, the sigma complex is of much higher energy than the starting material. This is mainly because the ring has lost its aromaticity, and the associated stability, at this point.

- Under conditions like this, the intermediate will immediately begin to look for ways to dispense with the positive charge of the carbocation and recover its aromaticity. The simplest way to remove the newly acquired positive charge is to lose a proton to the solvent.

- If we remove one of the substituents about the tetrahedral center that we generated in the first step of the reaction in a heterogenic bond cleavage, we can repopulate the aromatic ring using the bonding electrons that were holding that substituent in place. Removal of the electrophile would produce an uninteresting result, because we simply return to the starting materials. But what if instead we remove the proton on the same tetrahedral center?

- A molecule of solvent or another weak base is usually all it takes to abstract that proton, leading to a return to aromaticity (and therefore stability), while leaving the new benzene-electrophile bond intact. Our substitution reaction is complete.

Generating Electrophiles

- As simple as the electrophilic aromatic substitution mechanism seems, getting such a reaction started can be complicated because it requires a species so electronegative—so hungry for electrons—that it can compensate for the temporary loss of aromaticity during the reaction's first step. Indeed, the key to a successful substitution is to generate an adequately electrophilic substitution reagent.

- With regard to Zeidler's synthesis of DDT, chloral itself is not electrophilic enough to react with the aromatic electrons of chlorobenzene. This is why Zeigler added a small amount of sulfuric acid to his reaction. Sulfuric acid is strong enough to protonate a carbonyl group like that of chloral, increasing the electrophilic character of its carbonyl carbon. It is this protonated carbonyl species that was key to starting the reaction that ultimately produced this history-changing compound.

- Over the past two centuries, many researchers have devised ways to substitute benzene using many familiar functional groups, including nitration, halogenation, and alkylation. In all three of these cases, simply mixing benzene with nitric acid, molecular bromine, or an alkyl halide will not achieve the desired reaction. We need to find ways to make each of these three reagents even more electrophilic than they already are.

- Nitration of benzene is typically achieved by adding a bit of sulfuric acid to the mixture. The purpose of this sulfuric acid is to protonate the nitric acid, which acts as a base. Sulfuric acid is extremely strong, with a pK_a of about -2, and can protonate just enough nitric acid to get the process started.

- Once protonated, nitric acid can lose a water molecule to form a species known as nitronium. The electron cloud around nitronium is exactly the same as that of carbon dioxide, but it has a central nitrogen atom instead. That extra proton in the nucleus of nitrogen means that the nitrogen now has a +1 charge associated with it. This dense region of positive charge, complete with oxygen atoms ready to accept pi electrons as the attack occurs, is all it takes to make an effective electrophile for nitration.

- Halogenation of benzene is another highly desirable goal, but it is one that cannot be achieved simply by mixing benzene and molecular halogens like chlorine. In this case, inorganic complexes like aluminum(III) chloride or iron(III) chloride are used to induce a strong dipole in the molecular halogen bond. This method of chlorination is so effective that when taken to extremes, all six hydrogens from a benzene can be substituted with chlorine, producing hexachlorobenzene.

- The same type of catalyst can be used to enhance the dipole of an alkyl halide to produce an alkylation. Alkylated benzene compounds produced in this way can be used to produce such useful products as synthetic detergents.

Substitution of Substituted Benzenes

- When we substitute one nitrogen from benzene, like its nitration, or when we substitute all of its protons, like we did using chlorine and aluminum(III) chloride catalyst, only one possible isomer of the intended product exists. But what about when we replace some, but not all, of the hydrogens around a benzene ring?

- Once a substituent has been placed on the ring, the remaining hydrogens are no longer equivalent. We have two ortho positions, two meta positions, and one para position around the ring. Will the second substitution take place preferentially at one of these positions, or will we get a statistical mixture of all three possibilities?

- This consideration is well illustrated by Zeidler's successful DDT synthesis. If his electrophilic aromatic substitution methodology caused substitutions at random locations around the ring, we would expect that any given sample of chlorobenzene molecules should produce a mixture of products that are 40% ortho, 40% meta, and only 20% para.

- Moreover, the chances of a second substitution occurring at the para position are only 20%, leaving a paltry 4% of the final product in the preferred configuration. However, when Zeidler's synthesis is performed, the desired regioisomer is the major product obtained. So, how did it come to be that chlorobenzene selectively undergoes electrophilic substitution at its para position?

- This tendency for a reaction to occur at one particular position among several similar positions is called regiospecificity. Functional groups attached to benzene rings alter not only the reactivity of the ring, but also the regiospecificity of subsequent additions.

- Recall that there are three different positions available to react on a substituted benzene ring: two ortho positions, two meta positions, and one para position. Based on this distribution, one might expect that, for example, brominating a nitrobenzene molecule would yield a mixture of products that is 40% ortho-, 40% meta-, and 20% para-substituted.

- However, when we conduct this reaction, we find that this distribution of products is rarely the case! In fact, we usually find a preponderance of one product with just a small amount of the others. So, how does this preference to substitute at a particular location on the ring happen? The answer lies in the stability of the sigma complex transition states of all three reaction pathways.

- Let's start by analyzing the three possible pathways for the electrophilic aromatic substitution of chlorobenzene. When an electrophile finds itself at the meta position, we see a transition state form in which three resonance contributors form, all of which look very similar to those in a simple benzene ring undergoing substitution.

- However, when an electrophile attaches to the para position or the ortho position of the ring, we can draw a fourth resonance contributor. More contributors mean a more stable species. Hammond's postulate, named after chemist George S. Hammond, says that a more stable intermediate in the rate-limiting step leads to a faster reaction.

- In terms of the three possible products of any electrophilic substitution of chlorobenzene, the meta-intermediate is similar in energy to substitution of benzene, but the ortho- and para-intermediates are lower in energy, thus lowering the activation energies of the pathways for these two products.

- So, the meta position of chlorobenzene is less reactive than the ortho and para positions. In fact, chlorine is part of a class of substituents called ortho/para directors. To explain why the para position is substituted more than the ortho requires that we look to our second consideration: ring activation and deactivation.

- The second consideration is that the already-attached group may withdraw or donate electron density to the ring. Reducing electron density within the ring makes it less likely to attack, because doing so would only increase the magnitude of charge generated in the intermediate. Conversely, donating additional electron density to the ring would be expected to activate the ring and make it even more reactive.

- One of the traits that makes chlorine and bromine so interesting in this reaction type is that they withdraw electrons from the ring inductively. This means that the electronegativity of chlorine causes it to pull electrons out of the ring. Unlike resonance, inductive effects attenuate with distance.

- This gives us an explanation for the strong preference for the para position over the ortho position. The higher positive charge on the ortho atoms of chlorobenzene simply makes them less likely to attack an electrophile and take on even more positive charge.

Common Substituents and Their Effects

- Activating ortho/para directors include groups such as the amine group of aniline, the hydroxyl group of phenol, and the methoxy group of anisole. Groups like these have the ability to donate electrons to the ring by resonance.

- Deactivating ortho/para directors include halogens like fluorine, chlorine, and bromine. These groups direct by donating electrons in resonance but deactivate because of the strong dipole they create in the C-X bond.

- Deactivating meta directors, such as nitrates, carboxylic acids, and aldehydes, all withdraw electrons strongly by resonance. All of these groups contain a heteroatom two bonds away from the ring.

Suggested Reading

Casron, *Silent Spring*.

McMurry, *Fundamentals of Organic Chemistry*, Chaps. 5.3–5.7.

Wade, *Organic Chemistry*, Chaps. 17.1–17.3, 17.5–17.8.

Walker, *Blizzard of Glass*.

Questions to Consider

1. Is it possible for a benzene ring to undergo substitution with a nucleophile instead of an electrophile? If so, what structural features must the ring have (leaving groups and ring substituents)?

2. Propose a strategy to convert benzene into 1-bromo-2,4-dinitrobenzene using electrophilic aromatic substitution. What by-product do you expect to form in this synthesis?

Modifying Benzene—Aromatic Substitution
Lecture 21—Transcript

So last time we took a look at how benzene rings were first discovered, the debate over their structure, and exactly why they possess such an unusual stability. Now the benzene ring is a pervasive motif in organic chemistry not only because of its stability, but also for its complex, yet predictable and controllable reactivity. As we discussed previously, benzene is an aromatic compound. In fact, it's the structural basis for aromatic compounds, though, how they got this name is something of a mystery, since only some of them have strong or pleasing odors. Nonetheless, this moniker has stuck with benzene and its derivatives for nearly two centuries.

Now, as an aromatic, the pi electrons in a benzene ring are not nearly as susceptible to undergoing addition reactions as their non-aromatic counterparts. We discussed how heats of hydrogenation are evidence of just how stable benzene rings really are in the face of addition reactions, which require them to sacrifice their aromaticity. But what about substitution reactions? Is it possible to entice benzene to react in such a way that one of its six-ring hydrogens can be replaced with a more complex functional group of our choosing without sacrificing its aromaticity?

The answer to this question is yes. And one of the most frequently employed pathways to accomplish this is electrophilic aromatic substitution. During the late 1800s and early 1900s, the science of organic chemistry was booming in both Europe and America. Many names with which we are familiar today were hard at work during this time period devising new ways to make and break chemical bonds in a useful fashion. One such chemist was Adolf von Baeyer, who in 1873 was working at the University of Strasbourg in Germany. Now, Baeyer was supervising a graduate student by the name of Othmar Zeidler, whose graduate project included the synthesis of a substituted benzene compound known as dichlorodiphenyl trichloroethane—DDT for short.

Zeidler's synthesis of his compound was accomplished using chlorobenzene and chloral under acidic conditions. Under these conditions, the two chlorobenzene rings are joined by substituting the chloral for hydrogens in the para location on each ring. Zeidler's synthesis in 1873 was largely viewed as an academic exercise. Neither he nor his revered mentor, Baeyer, had any idea that this compound would ultimately alter the course of human history some seven decades later.

See, it was only during and after World War II that the insecticidal properties of DDT were discovered and applied with breathtakingly effective results. The Swiss chemist Paul Muller is credited with later discovering and documenting, and also obtaining Swiss patents for, the insecticidal use of this compound. After which, the U.S. government tested and used the compound to combat infectious diseases abroad. You see, DDT was such a powerful, broad-spectrum and long-lasting insecticide that its use could eradicate entire populations of disease-carrying insects so efficiently that the disease they carried was eliminated as well.

This includes a notable quelling of a typhus outbreak in Naples in the dead of winter. Until that time, such insect-borne epidemics could only be combated by bearing down and waiting for spring. Additionally, there are many examples in which DDT was used to wipe out malaria completely in confined populations. Shortly after the conclusion of World War II, the U.S. government began to allow commercial use of DDT for pest control in domestic crop production. Clearly, DDT had changed the world.

In the Nobel Prize awards ceremony speech for Dr. Muller, Professor G. Fischer noted of Muller's discovery quoting here, "At requisite insecticidal dosages, it is practically nontoxic to humans and acts in very small dosages on large number of various species of insect. Furthermore, it's cheap, easily manufactured, and exceedingly stable." Clearly, in 1948 the scientific community was riding high on the success of this newly discovered use of DDT.

Unfortunately, the stability and long-lasting effects of DDT also proved to be its undoing. DDT lasted so long in the environment that it could be transmitted from insects to fish and from fish to birds, where it had the unintended effect of inhibiting calcium production in those birds. This caused them to produce eggs with thinner shells than normal. And many bird species, most notably birds of prey, like eagles, osprey, and hawks, suffered tremendous population declines due to the presence of DDT in their food sources.

In 1962, environmentalist and Johns Hopkins University-trained marine biologist Rachel Carson published the book, Silent Spring. This book is considered one of the seminal works in modern environmentalism, and it spends a great deal of its time on the U.S. agricultural industry's use of DDT as a prime example of how rampant use of such compounds can lead to environmental disaster. It was only in 1972, just shy of a century after its first synthesis and three decades since its implementation as an insecticide, that the U.S. banned the use of DDT in agricultural applications. Today, it only finds use in emergencies as a vector control agent to prevent the spread of disease.

So why am I telling you this story? Well, because it's a shining example of how organic chemistry can change the world, both for the better and for the worse. The synthesis of DDT is also a shining example of an electrophilic aromatic substitution reaction, the topic of today's lesson. Today, we will build on our understanding of the structure of benzene by investigating this very useful class of reactions. We'll investigate the general mechanism by which it takes place. We'll discuss how other reagents often need to be activated to make them reactive enough to entice the substantially stabilized benzene ring to react with them. We'll learn about just a handful of the many modifications that we can make to benzene. And finally, we'll consider how already modified benzenes can be further modified to produce some of those ortho, meta, and para isomers that we talked about last time.

Electrophilic aromatic substitution is so named because the newly-attached motif starts out as an electrophile, unlike nucleophilic substitution, in which the newly added group is nucleophilic. So let's begin by looking at a reaction coordinate diagram for the reaction of benzene and a generic electrophile. Here, I'm just going to use E+ to identify the electrophile, but we'll soon look at some real examples.

Recall that even though they're in an aromatic system, the pi electrons of benzene are still not held as tightly as most sigma bonding electrons. And although it takes an extraordinarily electrophilic reagent to coax them into attacking, it can be done. When an adequately electrophilic reagent is added to benzene, the pi electrons of benzene simply can't resist attacking, and in doing so, they generate a tetrahedral center adjacent to a resonance-stabilized carbocation.

We call this intermediate a sigma complex, because we have traded two of our pi electrons to make a new sigma bond to the electrophile. Clearly, even with the newly-formed, more stable sigma bond, the sigma complex is a much higher energy than the starting material. And this is mainly because the ring has lost its aromaticity and the associated stability that comes from that aromaticity. Under conditions like this, the intermediate will immediately begin to look for ways to dispense with the positive charge of the carbocation and recover its aromaticity. Naturally, the simplest way to remove the newly-acquired positive charge is to lose a proton to the solvent. But which proton, exactly, will be lost?

Take a look at that tetrahedral center we generated in the first step. If we remove one of the substituent about this center in a heterogenic bond cleavage, we can repopulate the aromatic ring using the bonding electrons, which were holding that substituent in place. Removal of the electrophile would produce an uninteresting result, since we'd simply return to the starting materials. But what if instead we removed the proton on the same tetrahedral center? A molecule of solvent or another weak base is usually all it takes to abstract that proton, leading to a return to aromaticity, and therefore, stability while leaving the new benzene electrophile bond intact. Our substitution reaction is complete.

Zeidler's synthesis of DDT relied on this mechanism to couple two molecules of chlorobenzene to a compound known as chloral. He was able to substitute chloral for one of the ring hydrogens of chlorobenzene, creating an intermediate, which substituted for yet another ring hydrogen on a second molecule of chlorobenzene. This somewhat complicated use of electrophilic aromatic substitution will make more sense to us as we continue through the lecture.

As simple as our electrophilic aromatic substitution mechanism seems, getting such a reaction started can be complicated because it requires a species so electronegative, so hungry for electrons, that it can compensate for the temporary loss of aromaticity during the reaction's first step. Indeed, the key to successful substitution is to generate an adequately electrophilic substitution reagent. Now, remember what makes a good electrophile, high positive charge density and the ability to accept electrons to form a new bond.

Back to Zeidler's synthesis of DDT. Chloral itself is not electrophilic enough to react with the aromatic electrons of chlorobenzene, and this is why he added a small amount of sulfuric acid to his reaction. Sulfuric acid is strong enough to protonate a carbonyl group like that of chloral, increasing the electrophilic character of its carbonyl carbon. It's this protonated carbonyl species which is key to starting the reaction, which ultimately produced his history-changing compound.

Over the past two centuries, many researchers have devised ways to substitute benzene using many familiar functional groups. Now, three of these groups we'll discuss today. Those are nitration, halogenation, and alkylation. In all three of these cases, simply mixing benzene with, say, nitric acid, molecular bromine, or an alkyl halide will not achieve the desired reaction. We need to find ways to make each of these three reagents even more electrophilic than they already are. So let's look at how each of these three can be accomplished one by one.

Nitration of benzene is typically achieved by adding a bit of sulfuric acid to the mixture. The purpose of the sulfuric acid is to protonate the nitric acid. That's right, nitirc acid is going to act as a base, and this may sound crazy, but sulfuric acid is extremely strong with a pK_a of about -2, strong enough that it can protonate just a small amount of nitric acid and get the process started.

Once protonated, nitric acid can lose a water molecule to form a species known as nitronium. The electron cloud around the nitronium is similar to that of carbon dioxide, but it has a central nitrogen atom instead. Now, that extra proton in the nucleus of nitrogen means that the nitrogen now has a $+1$ charge associated with it. And this dense region of positive charge, complete with oxygen atoms ready to accept pi electrons as the attack occurs, is all that it takes to make an effective electrophile for nitration. Its reaction can be used to produce nitrobenzene, which is frequently used as an organic solvent, but it can also be used to create a number of very useful products. Let's take a look at one now.

Let's nitrate a compound called phenol using the same strategy that we just discussed. This is the structure for phenol. It's benzene with a hydroxyl group attached and is very easily obtained from coal tar extracts. So this compound was available to chemists working in the early 1800s. If we add to that sulfuric acid, H_2SO_4, with the sulfur atom indicated here in yellow, and nitric acid, HNO_3, with our nitrogen atom in the center here in blue, we can expect to generate a nitronium electrophile.

Now, all these materials have been readily available for quite some time, so this was a very simple reaction to run very early in the history of organic chemistry. Our sulfuric acid is going to transfer its proton to the nitric acid. Even if only a small amount does this, it's enough to get the reaction started. The product of this is our protonated nitric acid and a hydrogen sulfate anion, the conjugate base of H_2SO_4.

Now, once protonated, the nitric acid can lose this water functional group, generating the nitronium ion and a water by-product. As the reaction proceeds, the nitronium ion is attacked by the phenol and forms that new sigma bond to create the sigma complex. But we've lost our aromaticity now, and we have a carbocation instead. So to relieve this loss of aromaticity, in this case, I'm going to show water taking the proton from the tetrahedral center, which is in back here just behind the carbon, and in doing so, it will reestablish aromaticity, having left behind the newly-attached nitro group.

So we've completed one round of nitration of phenol. But it doesn't have to stop there. There are four more hydrogens on the ring. And in the case of nitronium and the case of phenol, they react so well together that using a little bit more nitric and sulfuric acid can result in this compound, a trinitrophenol, known as picric acid. Now, picric acid is noteworthy because it's a chemical cousin of trinitrotoluene or TNT. but, it was easier to make, was discovered much earlier, and believe it or not, has an even higher explosive yield.

No community can attest better to the explosive power of picric acid than the community of Halifax, Nova Scotia, which was devastated by an immense blast from a French cargo ship carrying this wartime munition in 1917. After colliding with another ship in the bay, the French ship, *Mont-Blanc* caught fire, prompting the crew to evacuate the ship. The resulting blast when the cargo detonated has been estimated at a force equivalent to about 3 kilotons of TNT, creating a one-mile-wide swath of complete and utter destruction.

Now, that yield is about $\frac{1}{5}$ the explosive yield of the nuclear bomb, which destroyed Hiroshima nearly 30 years later at the end of World War II. This is a true testament to the reactivity of nitro groups and what they can bring to organic compounds as the aromatic stabilization of benzene is clearly dramatically overwhelmed by the instability of the three nitro groups which are introduced to it by this method.

Halogenation of benzene is another useful synthetic process, but again, one which cannot be achieved simply by mixing benzene and molecular halogens, like chlorine. In this case, inorganic complexes, like aluminum(III)-chloride or iron(III)-chloride, are used to induce a strong dipole in the molecular halogen bond. This method of chlorination is so effective that when taken to extremes, all six hydrogens from a benzene can be substituted with chlorine, producing hexachlorobenzene.

Now, this halogenated hydrocarbon is a powerful anti-fungal agent, which was used in the 1940s, '50s, and '60s to protect wheat crops before its use was discontinued over, you guessed it, environmental concerns. The same type of catalyst can be used to enhance the dipole of an alkyl halide to produce an alkylation. Now, alkylated benzene compounds produced in this way can be used to make such useful products as synthetic detergents. And there's a long library of catalyst and tricks to attach a wide variety of functional groups to benzene, but we're going to conclude our discussion with these three examples. And as I continue through the remainder of the lecture, I'll try to use these three for demonstration purposes. But keep in mind that a huge range of modifications are possible.

So we've taken a look at what happens when we substitute just one hydrogen on a benzene ring, like its nitration, and when we substitute all of its protons, like we did using chlorine and aluminum three chloride catalyst. In cases like these, only one possible isomer of the intended product exists. But we also took a look at picric acid, which has some, but not all of the hydrogens around its aromatic ring substituted. And this presents a new consideration. The isomer that I showed you in the slide has three nitrates at the 2, 4, and 6 positions. But, why not this isomer, 2, 3, 4, or, maybe this isomer, 3, 4, 5?

Now you can see the issue that we're facing here. Once a substituent has been placed on the ring, the remaining hydrogens are no longer equivalent. We have two ortho, two meta, and one para positions around the ring. So we must now ask ourselves if the second substitution will take place preferentially at one of these positions, or, will we get a statistical mixture of all three possibilities. This consideration is well illustrated by Zeidler's successful DDT synthesis.

I say successful in the sense that he was able to modify two chlorobenzenes only at their para positions. If his electrophilic aromatic substitution methodology caused substitutions at random locations around the ring, we would expect that any given sample of chlorobenzene molecules should produce a mixture of products, which are 40% ortho, 40% meta, and only 20% para. Moreover, the chances of a second substitution occurring at the para position are, again, 20%, leaving a paltry 4% of the final product in the preferred configuration.

Yet when Zeidler's synthesis is performed, the desired regioisomer is the major product that's obtained. So how did it come to be that chlorobenzene selectively undergoes electrophilic substitution at its para position? We call this tendency for a reaction to occur at one particular position among several similar positions, regiospecificity. Functional groups attached to benzene rings alter not only the reactivity of the ring, but the regiospecificity of subsequent editions as well.

Now recall we noted earlier that there are three different positions available to react in a substituted benzene ring, two ortho, two meta, and one para, so based on this distribution, one might expect that, for example, brominating a nitrobenzene molecule would yield a mixture of products which is 40% ortho, 40% meta, and 20% para substitute. Yet, when we conduct this reaction, we find that this distribution of products is rarely the case. In fact, we usually find a preponderance of one product with just a small amount of the others.

So how does this preference to substitute at a particular location on the ring happen? Well, the answer lies in the stability of the sigma complex intermediates of all three reaction pathways. So let's start by analyzing the three possible pathways for the electrophilic aromatic substitution of a chlorobenzene molecule. Notice that when an electrophile finds itself at the meta position, we see an intermediate form in which resonance contributors can form, all of which look very similar to those in a simple benzene ring undergoing substitution.

However, when an electrophile attaches to the para position or the ortho position of the ring, something interesting happens. We can draw forth resonance contributor. And you may recall from our earlier discussions on resonance that more resonance contributors means a more stable species. And we can rely on something called Hammond's postulate, which tells us that a more stable intermediate in a rate-limiting step leads to a faster reaction. So, plotting reaction coordinate diagrams for the three possible products of any electrophilic substitution of chlorobenzene might look something like this. With the meta intermediate being similar in energy to the substitution of benzene but the ortho and para intermediates being lower in energy, thus, lowering the activation energies of the pathways for those two products.

So, we have successfully demonstrated why the meta position of chlorobenzene is less reactive than the ortho and the para positions. In fact, chlorine is part of a class of substituents which we call ortho-para directors. However, we've not yet explained exactly why the para position is substituted more than the ortho in this case.

Now, this requires that we look to our second consideration, ring activation and deactivation. See, the second consideration here is that the already-attached group may withdraw or donate electron density to the ring. Reducing electron density within the ring makes it less likely to attack, since doing so would only increase the magnitude of the charge generated on the intermediate. Conversely, donating additional electron density to the ring would be expected to activate the ring and make it even more reactive.

One of the traits which makes chlorine and bromine so interesting in this reaction type is that they withdraw electrons from the ring inductively. This means that the electronegativity of chlorine causes it to pull electrons out of the ring through bonds like current through a wire. And unlike resonance, which has the effect of causing charges to jump every other atom, inductive effects attenuate with distance. So this gives us an explanation for the strong preference for the para substitution over ortho. The higher positive charge on the ortho atoms of chlorobenzene simply makes them less likely to attack an electrophile and take on even more positive charge.

If we turn our attention to the synthesis of DDT one more time, we now understand how Zeidler was able to produce the specific isomer he wanted. The chlorine substituent on his starting material directed the chloral substitution to the para position, ensuring that the proper structural isomer was obtained. Not only that, but the newly-attached chloral group quickly dehydrates under acidic conditions to produce a new electrophile. This deactivates the already-attached ring, discouraging additional substitutions, while simultaneously encouraging a second chlorobenzene to attack, leading to the product, DDT. The final product of Zeidler's synthesis can be thought of as two disubstituted rings bearing deactivating substituents. So the desired product forms in great abundance with very limited by-product formation.

So, now that we understand how substituents can affect the sites on which substituted benzenes are modified in electrophilic aromatic substitutions, let's take a quick look at some common motifs and how they affect reactivity.

Activating ortho-para directors include groups, such as the amine group of aniline; the hydroxyl group of phenol; and the methoxy group of anisole. Groups like these have the ability to donate electrons to the ring by resonance. Deactivating ortho-para directors include halogens, like fluorine, chlorine, and bromine. These groups direct by donating electrons in resonance, but deactivate because of the strong dipole they create in the carbon-halogen bond. Finally, we have deactivating meta directors, such as nitrates, carboxylic acids, and aldehydes, which all draw electrons strongly by resonance.

Now notice that all of these groups contain a heteroatom two bonds away from the ring, while all of our activators contain the heteroatom adjacent to the ring. That's not a coincidence. Now the true utility and importance of directing effects on synthesis can be seen in this simple illustration. The bromination of nitrobenzene produces mainly meta bromo nitrobenzene, owing to the meta directing nitro group. Yet, the nitration of bromobenzene produces mainly para bromonitrobenzene, owing to the ortho-para-directing bromo group. So in this case, the order in which I carry out my substitutions on the benzene has a direct effect on the product that I obtain.

Now let's consider the activating affects of substituents and how they can affect a synthetic outcome as well. A one-to-one mixture of benzene and chlorine with a small amount of aluminum chloride catalysts should produce a quantitative, or complete conversion, to chlorobenzene, owing to the deactivating nature of the chloro groups. Since chlorobenzene itself is less reactive than benzene, we don't expect much of the chlorinated product, or rather, we don't expect much of the dichlorinated product to form.

Yet, a one-to-one mixture of benzene with ethyl chloride and catalytic aluminum chloride produces a mixture of benzene, ethylbenzene, diethylbenzene isomers, and even some triethylbenzene. Now, the reason for this is that the alkyl groups are activating, so as a small amount of ethylbenzene forms in the reaction mixture, its increased reactivity means that it's conversion to diethylbenzene will outpace alkylation of the remaining starting material. In situations like this, we have to turn to some tricks to achieve a single alkylation. For example, using a huge excess of benzene in order to dilute the electrophile to the point at which two alkylations on a single molecule becomes very unlikely.

Today, we have investigated the ability of electrophiles to substitute themselves for ring hydrogens of benzene molecules. We've discussed how the resonance stabilization of the resulting transition state makes it possible to entice a ring to give up its aromaticity, even if only for an instant while the modification is occurring. We've looked at ways to generate electrophiles, including acidification of substrates, like nitric acid and chloral, as well as the use of a Lewis acid catalyst, like aluminum three chloride, to activate halogens and alkyl halides.

And finally, we considered the effect of having a substituent already attached to the ring. We saw how substituents, like amine, hydroxyl, and methoxy groups, which donate electrons to the ring activate an ortho-para direct, how substituents like aldehydes and nitrates, which withdraw electrons from the ring by resonance, deactivate and meta direct. And we also considered how the high electronegativity of halogens gives them the very unusual property of deactivating while ortho-para directing.

And of course, along the way we considered how all of these factors combined in a simple, highly-efficient synthesis of a remarkably stable molecule, a molecule with such powerful biological activity that it not only changed the world, but ultimately, the way that humans think about how we interact with the world—powerful chemistry indeed.

Speaking of biological systems, in the next few lectures we're going to take our organic chemistry knowledge to the interface between chemistry and biology as we investigate the large, complex molecules of life, including carbohydrates, nucleic acids, and proteins through the eyes of an organic chemist. I'll see you then.

Sugars and Carbohydrates
Lecture 22

From the sweet ingredient of soda, to the cell walls making up a fibrous piece of wood, to the tough shell of a crab, all of the members of this extraordinarily diverse group of materials have one thing in common: They are all made up of the same general class of organic compound—carbohydrates. In this lecture, you will learn about these and other biologically important materials crafted from carbohydrates from the perspective of an organic chemist.

Definition and Classification of Carbohydrates

- Carbohydrates are compounds with the molecular formula $Cn(H_2O)$ n. Some simple examples of members of this class of compounds are the simple sugars erythrose, ribose, glucose, and fructose. Despite having a carbon-to-hydrogen-to-oxygen ratio of 1:2:1, carbohydrates are complex, polyhydroxylated species.

- Probably a more modern definition of carbohydrates would be polyhydroxyaldehydes and polyhydroxy ketones with the formula $Cn(H_2O)n$—and also the products produced by linking them together. Today, the term "carbohydrate" is used interchangeably with the terms "saccharide" and "sugar" to refer to all of these materials collectively.

© egal/iStock/Thinkstock.

We are trained to think of carbohydrates as not much more than a dietary source of calories, but they are actually much more than that.

- In 1861, Aleksandr Butlerov published a paper entitled "Formation of a Sugar-Like Substance," in which he detailed a synthesis using formaldehyde and a base to form what he believed to be simple sugars. He named his new substance "formose," for the formaldehyde starting material and the class of sugar that he believed he had created. In fact, Butlerov had created a mixture of carbohydrates that interconverted through a series of aldol reactions.

- Carbohydrates are broadly classified by the number of subunits that comprise them. Individual simple sugar molecules can be connected to one another in one of several ways to produce chains of varying length. These chains are still referred to as carbohydrates and are classified based on their length.

- Sugars that cannot be hydrolyzed into simpler compounds are called monosaccharides. When two such units are joined together in a condensation reaction, we get a disaccharide. Polysaccharides can have anywhere from just a few to thousands of condensed monosaccharide units.

Monosaccharides

- Monosaccharides can be of varying length and organization but must meet the requirement of having a molecular formula of $Cn(H_2O)n$. If we simply have a chain of carbons each with its own hydroxyl group, then we miss the required ratio of atoms just slightly—we have two hydrogens too many. To fix this, we reduce the compound to get a carbonyl at one end.

- The inclusion of just one carbonyl solves the problem and gives us a framework on which to build our simple sugars. When the requisite carbonyl is placed at the end of the chain, we call these sugars aldoses (for the aldehyde group). If instead the carbonyl is on an interior carbon, we call this a ketose (for the ketone motif it produces).

- We name monosaccharides based on the number of carbon atoms contained therein. So, a six-carbon sugar with a terminal carbonyl is an aldohexose, whereas a six-carbon monosaccharide with an interior carbonyl is a ketohexose. This system is applied to all commonly occurring monosaccharides, which tend to have anywhere from three to seven carbons, with most ketoses having the carbonyl on the second carbon.

- But simply defining the number of carbons in the chain and the location of the carbonyl bond in a simple sugar isn't enough to let us unambiguously identify carbohydrates. All along a chain of hexoses, pentoses, tetroses, and even the triose glyceraldehydes, there are one or more chiral centers. Particularly in the case of longer monosaccharaides like hexoses, chirality can generate a large number of potential stereoisomers.

- The drawing method known as Fischer projections is used to investigate stereoisomers. Fischer projections are created by drawing a compound so that all of the horizontally depicted bonds are coming out of the plane of the page and vertical bonds are falling back behind it.

- Herman Emil Fischer won his Nobel Prize in part for his characterization of the structure of glucose, and a large part of his technique for this was to use his systematic form of projection, which made all of the stereochemical relationships in these complex molecules evident.

- As a general rule, we draw monosaccharides in a Fischer projection with the terminal carbonyl at the top of the drawing. We can then number carbons of the chain from top to bottom to facilitate discussion.

Cyclic Hemiacetals and Glycosides

- A carbonyl activated by acid in the presence of an alcohol forms a motif called a hemiacetal. In the case of monosaccharides, we have both a carbonyl and an alcohol available to participate in such chemistry intramolecularly.

- When this reaction takes place between the carbonyl of a carbohydrate and its terminal hydroxyl group, a relatively stable ring of five or six atoms can sometimes form.

- We call these cyclic sugars furanose in the case of a five-membered ring and pyranose in the case of a six-membered ring, each named for the cyclic ether that makes up its backbone. The result is an equilibrium between the acyclic aldose or ketose form and the cyclic furanose or pyranose form.

- The anomeric C1 of pyranose and furanose molecules is a crucial site for modification in both the lab and in nature. Close inspection of glucopyranose reveals that the C1 is the only carbon bonded to a pair of electron-withdrawing oxygen atoms. Thus, it is the most electrophilic and most likely to participate in substitution reactions, such as the reaction between alpha-D-glucopyranose and acidified methanol.

- This mixture generates an equilibrium mixture of both the alpha- and beta-functionalized monosaccharide. When the C1 is functionalized, we call the product a glycoside, and the newly added group is called an aglycone.

- Glycosides turn up in a long list of biologically relevant materials, including salicin from willow bark used to treat headaches in ancient times. Glycosides even appear covalently bonded to specific protein side chains in a special class of enzymes called glycoproteins. There, they serve several functions, including protecting certain proteins from enzymatic degradation and enhancing their water solubility.

Disaccharides

- Although theoretically it is possible for any hydroxyl of one monosaccharide to form a glycosidic linkage with another, there are three arrangements that we principally see in nature. We distinguish these linkages by numbering the monosaccharide acting as the "aglycone" with prime notations ('). Following this system, 1,1' to 1,4' and 1,6' linkages are found in nature.

The starting material for beer is the disaccharide called **maltose**.

- A familiar disaccharide with a 1,4' linkage is maltose, which is formed by the glycosidic linkage of two glucose units through an alpha anomer. One glucose acts as the glycoside, bonding through its C1, and the other acts as the "aglycone," bonding through its C4'. This leaves the C1' to interconvert between enantiomers through an equilibrium process.

- Maltose is most familiar as the starting material for beer, made by the hydrolysis of starches in the first part of the process. Much of the maltose is consumed by the yeast that is used to brew the beer, but a small amount is retained in the beverage.

- Although they do exist in a few disaccharides, 1,6' linkages are less common. They are actually more relevant to polysaccharides. One example of this type of linkage is gentiobiose, a 1,6' disaccharide between two glucose units. By connecting to the aglycone at its 6' carbon, additional conformational flexibility is introduced, making a nice building block for branched polysaccharides.

- A prime example of a 1,1′ linkage is the familiar sweetening agent sucrose, which consists of a glucose and a fructose unit linked by a bridging oxygen through one another's anomeric carbon. So, in this unusual case, both sugars are acting as glycosides for themselves and as aglycones for their bonding partner.

- Because both anomeric carbons are occupied by the glycosidic linkage, sucrose has no free anomeric carbon to open back up in equilibrium with the cyclic form. This is important because sugars are much more prone to oxidative degradation when in their acyclic forms. This makes sucrose much slower to oxidize than other disaccharides and a popular choice for use in preserves.

Polysaccharides

- Polysaccharides are simply longer chains of sugars in which most of them are acting both as glycosides and as aglycones, perpetuating a polymer built from monosaccharide subunits. Many materials with which we are familiar are made from polysaccharides. For example, cellulose in cotton fibers is made up almost exclusively of D-glucose units attached by beta-1,4′-glycosidic bonds.

- The same material makes up about half of the mass of dried wood. Cellulose combines into closely packed fibrils that interact with one another via hydrogen bonding, forming a strong, tough, insoluble material. Certain animal species, most notably cows and termites, have symbiotic bacteria living in their digestive tract that help them break down the cellulose in cotton and lumber and use them as a food source.

- By contrast, switching the beta linkage to an alpha linkage gives us amylose, a starch that curves on itself into a helix. It can be digested and used as a source of nutrition for higher-order animals—even those without bacterial help to break it down.

- If we take amylose and branch it about every 30 subunits or by cross-linking another chain at a 6′ position, we get amylopectin, a water-soluble thickening agent used in cooking applications.

- More extensive branching of the same kind leads to glycogen, one of the human body's primary energy stores. If you have enough glycogen stored up in your muscles, you are just a hydrolysis reaction away from additional glucose molecules for use as energy to help you run that marathon or finish that workday.

- A final example of polysaccharides is the remarkable armor that protects insects and arthropods such as cicadas and crabs. Chitin—the material from which these exoskeletons are made—is a polymer of a glucose derivative called N-glucosamine, which is a pyranose form of glucose with an amide functional group substituted at the C2 position of the ring.

- Polymerize this molecule through beta-1,4′-glycosidic bonds and you have chitin, a material so tough that once formed, it cannot grow with the animal. This is why crabs and insects periodically molt, shedding their chitin armor and producing a whole new, slightly larger coat of armor once the old one is shuffled off.

© SINGTO2/iStock/Thinkstock.

The same class of materials used to sweeten soft drinks can also make up the armor that protects insects and sea creatures from predators.

- The secret to chitin's strength is not only in the glycosidic linkages, but also in the amide functionality added to the sugar ring. The amide nitrogen and carbonyl from one chitin chain can hydrogen bond to others on a separate strand. The result of this H-bond network is a material of remarkable strength.

Suggested Reading

McMurry, *Fundamentals of Organic Chemistry*, Chap. 14.

Wade, *Organic Chemistry*, Chaps. 23.1–23.7, 23.18, 23.19.

Questions to Consider

1. Is cyclization of carbohydrates driven by enthalpy or entropy? In light of this, how is temperature expected to affect the equilibrium between open-chain and cyclized forms?

2. Draw the two chlorinated monosaccharide subunits comprising sucralose (discussed in Lecture 11), indicating stereochemistry where appropriate.

Sugars and Carbohydrates
Lecture 22—Transcript

In 1815, one of the largest volcanic eruptions in human history took place at Mount Tambora in Indonesia. This colossal volcanic eruption ejected an unfathomable amount of volcanic ash and debris into the atmosphere, blocking out a significant fraction of the sunlight reaching the Earth's surface. And this sent parts of the world into a chill from which they did not awaken until the following year.

In the middle of the year of 1816, this cloud drifted over Western Europe, leading to what is sometimes referred to as the year without a summer, which brought a devastating famine. Now, the result of this was a sustenance crisis, the likes of which the Western world has not seen in the 200 years since.

Particularly hard hit in this crisis was the nation of Germany, where food riots, demonstrations, and looting were commonplace that year. Now, one can only imagine how suddenly and incomprehensibly and difficult life must have become for the people of Europe as this crisis played itself out. One of those people was a 13-year-old boy by the name of Justus von Liebig Many historians believe that his traumatic event taking place at such a formative age so shook this young man's faith in mother nature to provide, that he dedicated his life later to researching agricultural chemistry.

Now, his motivations aside, it's undeniable that Liebig's contributions to the field of chemistry in the decades following the year without a summer opened up a whole new world of chemicals to study, specifically those chemicals involved in the biology of plants. Liebig earned his doctorate in 1822 at the University of Erlangen in Bavaria. After that, in his laboratory in Giessen, Germany, and later Munich, Liebig spent much of his career characterizing and synthesizing a broad range of biologically relevant compounds from plants. So prolific was his work that he's often cited as the father of agricultural chemistry, a well deserved acknowledgement.

Today, we will swing a pretty wide turn in our study of organic chemistry. We're going to move from the realm of the harsh reaction conditions and unfamiliar materials of our synthesis lectures into a more biologically oriented look at organic chemistry. And we're going to start with one of the classes of compounds that Liebig himself studied. Now, from the sweet ingredient of sodas, like this sugar, to the fiber particles making up this wood, to the tough shell of crabs and lobsters, all the members of this extraordinarily diverse family of materials have one thing in common—they're all made up of the same general class of organic compounds, carbohydrates.

In modern culture, we're trained to think of carbohydrates as not much more than a dietary source of calories. But we're about to see that carbohydrates make up not only small molecules of caloric value, but also other familiar materials, like clothing, building materials, joint supplements, and even body armor. In short, we'll take a look at these and other biologically important materials crafted from carbohydrates from the perspective of an organic chemist.

Now, before we break into the exciting and complex world of materials fashioned from carbohydrates, let's begin by defining a carbohydrate. This term is sometimes confused by students with the similar-sounding hydrocarbons that we encountered earlier in our course. Now, the distinction here is that the hydro of hydrocarbons refers to just hydrogen, while the hydrate of carbohydrates refers not just to hydrogen, but to water, which includes one half molar equivalent of oxygen.

So, as I mentioned before, carbohydrate refers to a compound composed of carbon and water in equal parts. So they'll have a generic formula of CnH_2On. Now, we've come pretty close to carbohydrates before with some of the compounds that we've looked at previously. Take the example of glycerol; now, glycerol's formula is $C_3H_8O_3$. So it's got a little bit too much hydrogen in the formula to meet our requirement to be classified as a carbohydrate. But I can modify my glycerol structure slightly to make a carbohydrate.

So let's bring up two copies of glycerol here and modify them to create a carbohydrate. I've got to get rid of two hydrogens. So let's start by removing two hydrogens from the end of my glycerol molecule. When I do this, I can insert a carbon-oxygen double bond creating $C_3H_6O_3$, a carbohydrate. But I can also remove two hydrogens from the center. If I remove two hydrogens from an interior motif here I get $C_3H_6O_3$, also a carbohydrate. But I changed the names to glyceraldahyde and glycerone in my two structures. Glyceraldahyde because the carbonyl is in a terminal position like it would be in an aldehyde. Glycerone because the carbonyl is in an interior carbon, just as it would be if these groups were aliphatic and I were dealing with a ketone. So what I've done here is created three different classes of compounds. Glycerol would be considered a sugar alcohol, whereas my glyceraldehyde and glycerone are sugars, which we call individually aldoses and ketoses, referring to whether the carbonyl is internal or external.

Now, I've mentioned that Alexander Butlerov before, most notably as the mentor of two of the most revered and mutually repulsed chemists of the 19th century, Markovnikov and Zaitsev. Now just as he's contributed to many other fields, the history of carbohydrate chemistry would not be the same without him. In 1861, Butlerov published a paper entitled "Formation of a

Sugar-Like Substance" in which he detailed a synthesis using formaldehyde and a base to form what he believed to be simple sugars. He named his new substance formose for the formaldehyde starting material he used and the class of compounds which he believed he created, sugars. In a sense, Butlerov was right. But he created not just one sugar, but a mixture of sugars, which interconverted through a series of aldol reactions forming a concoction of carbohydrates. Now in Butlerov's day, a quick taste would be all it took to make his observations publication worthy. Today, of course, we cringe at the idea of tasting anything prepared in the lab, and we require a bit more detailed analysis to identify the structures of the compounds that we make.

Butlerov wasn't able to surmise that there are many, many carbohydrates which can be produced and isolated from one another using techniques more modern to the science of organic chemistry. These carbohydrates are broadly classified by the number of sugar molecules which we combine to make them up. See, an individual simple-sugar molecule can be connected to another one in several ways which produce chains of varying lengths and geometries. And these larger structures built of simple sugars are still referred to as carbohydrates, but we call them complex carbohydrates, and these are classified based on their lengths and geometries.

But we're getting ahead of ourselves. Let's step back and consider the most fundamental of carbohydrates first. There are those which cannot be hydrolyzed into simpler carbohydrates, and those are the ones we want to deal with now. We call these kinds of sugars monosacharides. When two such units are joined together in a condensation reaction, we get a disacharide. And finally we have polysaccharides, which can have anywhere from just a few to literally thousands of condensed monosaccharide units.

So let's start at the beginning, the simplest class of carbohydrates—monosacchardies. As I mentioned before, we break down these simple sugars into two major classes, aldoses and ketoses. I've prepared the structures are four different aldoses for you here so we can see their structural similarities and differences. Now, the striking structural similarity here, of course, is that we have an aldehyde carbonyl. So we call this molecule an aldotriose, aldo indicating that we have a carbonyl at the end of the chain; triose indicating that there are three carbons in the chain. So, similarly, we would call this one aldotetrose, aldo for a terminal carbonyl, tetrose for four carbons in the chain. And the trend continues through aldopentose, and aldohexoses, and beyond.

Now we use a very similar naming system for ketoses, that is, when the carbonyl bond is interior. For example, this molecule would be a keto, because the carbonyl bond is internal, and a triose, because there are three carbons. So I have a ketotriose. My next example would be a ketotetrose, and so on through the remainder, a ketopentose and a ketohexose.

But, defining the number of carbons in the chain and the location of the carbonyl bond in a simple sugar isn't enough to let us unambiguously identify carbohydrates. You see, all along a chain of hexoses, pentoses, tetroses, and even triose, there are one or more chiral centers. Now, particularly in the case of longer monosacchardies, like hexoses, chirality can generate a large number of potential stereoisomers. Now, that's why I drew my sample molecules in the previous slides in a Fischer projection. Remember, a Fischer projection is created by drawing a compound so that all of the horizontally depicted bonds are coming out of the plane and all the vertical ones are falling back behind the plane. Now, Fischer won his Nobel Prize in part for his characterization of the structure of glucose, and a large part of this technique for this was to use his systematic form of projection, which made all of the stereo-chemical relationships in these complex molecules evident.

So let's begin with the simplest of all carbohydrates, the trioseglyceraldehyde. You can see clearly that glyceraldehyde has only one chiral center, so this will be a relatively easy puzzle to solve. As a general rule, we draw monosacchardies in a Fischer projection with the terminal carbonyl at the top and then going downward. So we can then number the carbons of the chain from top to bottom to facilitate the discussion.

Now, notice that only C2 of glyceraldehyde is chiral. So I can draw two enantiomers, one of which was named plus glyceraldehyde and the high and the other minus glyceraldehyde. Now remember, this was a long time before the Cahn-Ingold-Prelog rules had been proposed, and Fischer had only physical properties of saccharides to go on for identifying their stereochemistry. So, he designated glyceraldehydes as plus and minus based on their ability to rotate a special kind of light called plane-polarized light.

We're going to discuss this phenomenon in detail during our spectroscopy lectures. But for now, let's just say that the plus and minus glyceraldehyde designations stand for clockwise and counterclockwise rotation of that light. Now, if I start with glyceraldehyde and build larger monosaccharides by adding new interior carbons to the chain, I'll call my construct a D sugar. If, instead, I start with a minus glyceraldehyde, I'll have to call that construct an L sugar.

The D and L designations are, again, based on the molecule's interaction with light, D for clockwise from *dextra*, and L for counterclockwise from *levra*. For example, I can create D erythrose by inserting one new carbon into the chain of D glyceraldehyde. I could also make a molecule of D threose in the same way by changing the chirality of my newly inserted chiral center. So, not surprisingly, the introduction of a second chiral center has led to the possibility of diastereomers for my aldotetroses. Of course, I could, instead, have an L configuration for the bottom chiral carbon of my Fischer projection, leading to the other two possible stereochemistries, L aldotetroses, each a mirror image of one of the D aldotetroses.

Now this interesting set of stereoisomers in the Fischer projection combined to produce a new relationship in stereochemistry known as erytho and threo enantiomers. When aligned in a Fischer projection orientation, if similar groups are on the same side, as they are in erythos, then we use a prefix, erytho, to name that enantiomer. Conversely, if groups of interest are at opposition, we use the threo designation. As the size of the monosaccharide grows, so do the number of possible stereoisomers. By the time we reach hexoses, there are eight stereoisomers each for aldohexoses and for ketoses. Things get out of hand pretty quickly here.

Now, fortunately, there are just a handful of these which tend to make appearances in natural systems. And two of the most frequent are the aldohexose, known as glucose, and the ketohexose, known as fructose. For simplicity's sake, we'll be using just these two monosaccharides as examples as we go forward in this lecture. But keep in mind that the chemistry we're about to see can easily be applied to any of the dozens of small saccharides that can be created with just a few carbons in the chain.

So far, I've drawn all the carbohydrates under investigation today as linear compounds. But you may recall from our lecture on carbonyl compounds, that a carbonyl activated by acid in the presence of an alcohol forms a motif called a hemiacetal. Well, in the case of monosaccharides, we have both a carbonyl and an alcohol available to participate in this kind of chemistry intramolecularly. When this reaction takes place between the carbonyl of a carbohydrate and one of its own hydroxyl groups, a relatively stable ring of five or six atoms can sometimes form. We call these cyclic sugars, furanose in the case of a five-membered ring, and pyranose in the case of a six-membered ring. Each of these is named for the cyclic ether, which makes up their backbone. The result of this is an equilibrium interconversion between the acyclic aldose or ketose form and the cyclic furanose or pyranose form.

So let's take a look again at glucose, an aldohexose with a Fischer projection which looks like this. Now we see yet another reason why Fischer projections are so useful. Since all of the carbon-carbon bonds are already aligned so that they curve in on each other, the cis-trans relationships of groups within the cyclic form are very easy to determine from this projection. I simply turn my Fischer projection on its side to get a better look at the forming ring.

For example, hydroxyls which are erythro to one another in the aldohexose will be cis to one another in the resulting furanose. If I complete the cyclization in this fictitious orientation, I get what's commonly known as a Hayworth projection, with a planar six-membered ring, which we all know can't exist. Yet, we do this anyway sometimes because it so clearly shows the cis-trans relationships in the cyclic form of the sugar.

Now, once the ring is placed into a proper chair confirmation, we see that there's still one stereo-chemical consideration left. Since the carbonyl is planar, the attack on that motif during cyclization can take place from either face, meaning that there are really two different enantiomers of any pyranose or furanose sugar, which we sometimes call anomers, and we designate these as alpha and beta anomers.

In the case of D glucopyranose, all the existing hydroxyl groups can be placed in the equatorial position simultaneously. So it's not much of a wonder why this sugar likes to exist as a pyranose ring, right? If the newly formed hydroxyl is in the axial position, we call this an alpha anomer. If the new group is, instead, in the equatorial confirmation, we have the beta anomer.

And by convention, we assign the designation of C1 for the terminal carbon closest to the carbonyl end of the uncyclized carbohydrate and number each carbon from their sequentially. So the anomeric C1 glucopyranose is a crucial site for modification in both the lab and in nature. Close inspection of glucopyranose reveals that the C1 is the only carbon bonded to a pair of electron-withdrawing oxygen atoms. So, it's the most electrophilic and most likely to participate in substitution reactions, like this very simple reaction between alpha D glucopyranose and acidified methanol. This reaction generates an equilibrium mixture of both the alpha and beta functionalized monosaccharide. When the C1 is modified this way, we call the product a glycocide and the newly added group an aglycone.

So let's take a look at a glycocide that has some historical relevance. This is a molecule called salicin, and salicin is a glycocide with anti-inflammatory properties, and it's also present in the bark of willow trees. Ancient Greeks knew that chewing on the bark of a willow tree gave pain relief and alleviated headaches and joint problems. What they didn't know was that what they were doing while chewing the bark was extracting this particular compound, salicin.

Now salicin consists of a glucopyranose, which is bonded to a hydroxybenzyl alcohol, aglycone. And once this compound has been ingested, the water in your bloodstream and the enzymes in your blood go at breaking this down through a hydrolysis reaction. It takes a water molecule to break that link, separating the glucopyranose from the hydroxybenzyl alcohol. Now, at this point, the hydroxybenzyl alcohol is oxidized by other enzymes in your body in preparation for removing it. And in doing so it creates salicylic acid, the same compound into which aspirin degrades to give joint pain relief, headache relief, and other beneficial properties as well. Now, all of this amazing and rich chemistry is done with a few monosaccharides and an endless library of aglycones. But I believe that you were promised more. I still owe you an explanation of disaccharides and polysaccharides. So let's get started on those.

Our last topic was a perfect lead-in to disaccharides. We looked at how glycocidic linkages can form to almost anything with an amine or alcohol group on it, really, anything that's nucleophilic. And, if you're really on top of your game, you might have wondered if all those other alcohol groups on the sugar might make it possible for one sugar molecule to bond to another using the same motif. Now, if you had that thought, give yourself a gold star. You are spot on.

Although, theoretically it's possible for any hydroxyl of one monosaccharide to form a glycoacidic linkage with another, there are three arrangements which we principally see in nature. And we distinguish these linkages by numbering the monosaccharide acting as the aglycone with prime numbers. So check this out, take the system 1, 1', 1,4', and 1,6'. These are the linkages that we most often see in nature, meaning that the C1 will link to a different oxygen from one of the carbons in the other saccharide.

Now, a familiar disaccharide with this 1,4' linkage is maltose. Maltose is formed by the glycoacidic linkage of two glucose units through an alpha anomer. Now, here one glucose acts as the glycocide, bonding through its C1, and the other acts as the aglycone, bonding through its C4'. Notice that this leaves the C1' to interconvert between enantiomers through an equilibrium process. So, this disaccharide can still open and close in at least one of its rings.

Maltose is most familiar as the starting material for beer, made by the hydrolysis of starches from grains in the first part of the brewing process. Much of the maltose is consumed by the yeast, which are used to brew the beer. But a small amount is retained in the beverage. So the next time you have a beer with your friends, take a sip, and pronounce, that has some lovely 4O alpha D glucopyranocil D glucopyranose notes. You may wind up drinking alone after that, but at least you were telling the truth.

Now, the 1,6' linkages are less common, but they do exist in a few disaccharides. They're actually more relevant to polysaccharides as we'll soon see. One example of this type of linkage in a disaccharide is gentiobiose, a 1,6' disaccharide between two glucose units. Now, by connecting to the aglycone at its 6' carbon, which is not part of the pyranose ring, additional conformational flexibility is introduced, making a nice building block for branched polysaccharides.

Now a prime example of a 1,1' linkage is the familiar sweetening agent sucrose. Sucrose consists of a glucose and a fructose unit linked by bridging oxygen through one another's anomeric carbons. So in this unusual case, both sugars are acting as glycocides for themselves and as aglycones for their bonding partner. Since both anomeric carbons are occupied by the glycoacidic linkage, sucrose has no free anomeric carbon to open back up in equilibrium with the cyclic form. Now this is important because sugars are much more prone to oxidative degradation when in their acyclic forms. So, this makes sucrose much slower to oxidize than other disaccharides, and therefore, a popular choice for use in preserves. Polysaccharides, as you might expect, are simply longer chains of sugars in which most of them are acting as both glycocides and as aglycones, perpetuating a polymer built from monosaccharide sub units.

Many materials with which we're familiar are made from polysaccharide. For example, cellulose in the cotton fibers in these cotton balls is all made up exclusively of D glucose units attached by beta 1,4' glycoacidic bonds. The same material makes up about one half of the mass of dried wood. Now, cellulose combines into closely packed fibriles, which interact with one another via hydrogen bonding forming a strong, tough, insoluble material.

Now, cotton balls and lumber don't seem terribly appetizing to me as a human. But certain animal species, most notably cows and termites, have a symbiotic bacteria living in their digestive system which help them to break down the cellulose in these materials and use them as a food source.

Now by contrast, switching the beta linkage to an alpha linkage gives us amylose, a starch which curves in on itself into a helix. It can be digested and used as a source of nutrition for higher-order animals, even those without the bacteria to help them break it down. Now, if we take amylose and branch it about every 30 sub units or so by cross linking another chain at that 6' position, we get amylopectin, a water-soluble thickening agent used in cooking applications. More extensive branching of the same kind of leads to glycogen, one of the human body's primary energy stores. Now, if you have enough glycogen stored up in your muscles, you are just a hydrolysis reaction away from additional glucose molecules for use as energy, to help you run that marathon or finish that work day.

As a final example of polysaccharides, let's take a look at the remarkable armor, which protects insects or arthropods from ciccadas to crabs. You might find it hard to believe that the same class of materials that's used to sweeten soft drinks can also make up the armor which protects sea creatures and insects from predators. But it does. Kyton, the material from which this tough exoskeleton is made is a polymer of a glucose derivative called n glucosomine; n glucosomine is a pyranose form of glucose within an amide functional group substituted the at the C2 position of the ring. Now, polymerize this molecule through beta 1,4' glycoacidic bonds, and you have kyton, a material so tough, that once formed, it can't grow with the animal. And this is why crabs and insects periodically molt, shedding their kyton armor and producing a whole new, slightly larger suit of armor once the old one has been shuffled off.

The secret to kyton's strength is not only in the glycoacidic linkages, but also in that amide functionality added to the sugar ring. You can see that when the multiple strands of kyton are aligned, the amide, nitrogen, and carbonyl from one kyton can hydrogen bond to others on a separate strand. This h-ond network contributes to the remarkable strength of kyton, which protects against predation. It's also the reason that crustaceans and insects must molt, shedding their old skeletons during periods of concerted growth, during which a new exoskeleton is formed. The toughening provided by this hydrogen bonding network can simply be turned off again by modifying the amine group of glucosomine.

In the body, our joint cartilage is made up partially of glucose amino glycans, which are long polysaccharides consisting of alternating acetyl glucosomine and another modified sugar called uronic acid. As opposed to tough, inflexible kyton, glucosomino glycans like this one are springy and lubricating, because the hydrogen bonding network is disrupted by the acetyl motif, and this allows one chain to slide along another more easily. And that's the reason that many believers religiously take doses of glucosomine in the hopes of limiting cartilage loss associated with arthritis.

So let's sum up for the day. Today we discussed carbohydrates, a broad and diverse class of compounds with the chemical formula CnH_2On. We took a look at Fischer projections of the two main classes of simple carbohydrates, those being aldoses and ketoses, which differ by the location of there carbonyl motif.

We then discussed the stereochemistry of simple carbohydrates and the concept of erytho and threo enantiomers. After that, we discussed the cyclization of simple sugars by reacting a hydroxyl with the carbonyl in an intramolecular reaction to form a hemiacetal, and how the most common of these are five-membered furanose and six-membered pyranose rings. We also recognized that the hemiacetal carbon can have two distinct stereochemistries, leading us to label it an anomeric carbon.

We looked at how these cyclic sugars can react through their anomeric carbon to form alpha and beta glycosidic linkages to new groups called aglycones. When the aglycone is another sugar itself we form a disaccharide, like maltose from beer, or sucrose from fruit preserves. And we discussed the 1,4′, 1,6′, and 1,1′ glycoacidic linkages. And how the 1,4′ linkage is well suited to formation of long polymer strands. The 1,6′ linkage imparts branching in such polymers. And the 1,1′ linkage in sucrose prohibits its two sugar units from opening and oxidizing, making it a perfect choice for food applications because of its resistance to oxidation.

Finally we considered polysaccharides of glycopyranose, including cellulose, which has beta glycoacidic linkages, giving it tensile strength, in contrast to amylose and amylopectin, which have alpha glycoacidic linkages, endowing them with a higher solubility and making them more digestible for humans. We also saw how some animals, like arthropods and insects, chemically modify the glucocide subunits of cellulose to form even more powerful sets of interstrand hydrogen bonds, which endow their skeletons with formidable physical properties.

Next time, we're going to take a look at a very special aldopentose—ribose—and how it acts as the central piece of one of nature's most awesome molecular creations—DNA. I'll see you then.

DNA and Nucleic Acids
Lecture 23

E legantly simple in its library of just four letters, but stunningly complex in the size and organization of those four letters, DNA and its chemical cousin RNA are familiar biopolymers that form the basis of life itself. Our understanding of DNA and the biochemical processes surrounding it has altered our perception of what it is to be a living thing. Scientists endeavor to study it, understand it, and ultimately manipulate it in ways that improve life. None of this understanding is possible without tearing this beautiful complex down into its most basic constituent parts—which is the focus of this lecture.

RNA

- Ribonucleic acid (RNA) serves several critical roles in converting your genetic code, or genome, into the proteins essential to life, or proteome. In addition to other roles, one of RNA's most important jobs is to transport the information in the DNA of your cells to areas where proteins are made. The "ribo-" part of the name refers to the carbohydrate ribose.

- Like any other aldopentose, ribose can undergo cyclization to form the associated ribofuranose, with its anomeric carbon in one of two positions: alpha or beta. It is the beta form that is the central building block of RNA.

- To produce the building blocks of RNA, the anomeric hydroxyl in beta-ribofuranose is replaced with an organic base through a nitrogen atom. In the case of RNA, these four bases are cytosine, uracil (as opposed to the thymine of DNA), adenine, and guanine. These units are further subdivided into the class of pyrimidine bases and purine bases, based on their backbone structure.

- Using one of these four bases as the aglycone in beta-ribofuranose produces one of the four subunits of RNA, collectively called nucleosides. Not surprisingly, these nucleosides have names similar to those of the bases that give them their identity. An aglycone of cytosine forms the nucleoside cytidine. Uracil leads to uridine. Adenine produces adenosine, and guanine produces guanosine.

- The glue that holds the nucleoside subunits together into a chain— so that there is a sequence to the collection of C's, U's, A's, and G's that will make up the information in the RNA molecule—is a phosphate functional group.

- By attaching a phosphate group at the $5'$ position on the sugar of our nucleoside, we create a motif called a phosphoester. The phosphoester group will be the linkage through which our nucleosides are connected to give our collection of information order.

- This collection of three crucial elements—information in the base, the backbone of the ribose, and the glue of the phosphoester—make up what is called a ribonucleotide, which is one complete subunit of RNA, ready to be linked to others to create a useful code.

- RNA is nature's messenger molecule, transporting information from the genetic code stored in DNA and assisting in the conversion of that information into the vast library of proteins that our bodies use to regulate its chemistry.

- Because the role of RNA is transmission of information, which is carried out on a fairly rapid time scale, RNA itself does not need to endure for long periods of time to effectively perform its duties in living systems. In fact, it is part of RNA's design that it can be broken down easily and its components recycled for use in creating more RNA as it is needed.

- RNA has a pretty stable backbone, but there is a chink in the armor at the 2′ hydroxyl group. This nucleophile hangs out next to the phosphate ester group, which has an electrophilic phosphorus atom. The 2′ hydroxyl of RNA just can't help but attack the phosphorus, initiating a mechanism that ultimately expels the next ribonucleotide in the sequence, breaking the chain that gives RNA its information-carrying ability.

- RNA can carry huge amounts of information back and forth, being broken down and rebuilt as the organism needs it. But there would be no information to transport or translate if there were no method of storing that information indefinitely until it is called upon. This task falls to a close chemical cousin of RNA, the famous DNA.

DNA

- Deoxyribonucleic acid (DNA) is another ribose-based biopolymer that serves a very different purpose. DNA is very similar to RNA, consisting of a phosphate, sugar, and base to form what we call a deoxyribonucleotide.

- There are two key structural differences that make DNA more robust than RNA. First is the removal of the 2′ hydroxyl group of the beta-ribofuranose sugar. Instead, this group is replaced with a hydrogen, which is non-nucleophilic. This removal of one very specific oxygen from the ribose is the source of the "deoxyribo-" portion of DNA's name.

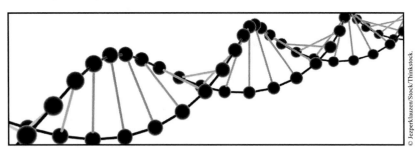

© Jezperklauzen/iStock/Thinkstock.

Mapping genes and sequencing DNA helps us understand the chemical basis on which life evolves and changes in response to environmental stimuli.

- With this nucleophile out of the picture, DNA hydrolyzes about 100 times slower than RNA. This makes DNA a much better choice not for a short-lived messenger, but for the storage of the genetic code that needs to last you the rest of your life.

- A second feature of DNA that makes it a more robust system is the replacement of uracil with thymine, a methylated version of uracil. The reason for this difference is that cytosine itself can deaminate spontaneously to become uracil. When this happens to your DNA, this is bad news, because it means that your DNA mutated.

Cytosine   Uracil   Cytosine

- But your body has a defense mechanism for this—a complex molecule called an enzyme that finds that uracil in your DNA and re-aminates it to recover the original cytosine. This incredible repair mechanism comes with a problem, however. If DNA actually used uracil, it wouldn't know which uracils to alter and which ones to leave alone. So, it uses thymine instead as the complementary base for cytosine.

Base Pairing
- DNA's double-helical structure was first proposed by James Watson and Francis Crick. But the discovery of the DNA double helix was the result of a concerted effort on the part of many talented researchers.

- In 1869, Friedrich Miescher first isolated DNA from the nuclei of cells. It was this biological source that prompted him to name it nuclein—a name that was later revised to DNA to better reflect its chemical composition.

- Russian researcher Phoebus Levene was able to identify the individual components of DNA, including the sugars, phosphates, and bases that comprised it. But he lacked the necessary tools to properly explain exactly how each of these pieces came together to form a structure for DNA.

- In the 1940s, Erwin Chargaff developed a method for isolating the bases from DNA samples. He found that the number of cytosine bases in any given sample closely matched the number of guanine bases. He also determined that the number of adenine bases closely tracked that of thymine.

- In the 1950s, the X-ray crystallography lab of Maurice Wilkins and Rosalind Franklin at Kings College finally reached a milestone in genetics. They were able to use a technique called X-ray diffraction to determine the spatial arrangements of atoms in the DNA molecule.

- In the end, it was Watson and Crick, however, who made it to press first with their model of the double helix, earning them their place in history. What Watson and Crick proposed explained every observation to date: The sugar, phosphate, and base molecules found by Levene were all accounted for; the equal amounts of purine and pyrimidine bases from Chargaff's experiments were easily explained; and the crystallography data from Wilhelm and Franklin's labs, as well as Watson and Crick's, clearly showed that a helical structure held them all together.

- The key to putting it all together was that the two pyrimidine bases (adenine and thymine), when held in place by the helix proposed, formed a network of two hydrogen bonds, causing them to pair up. Guanine and cytosine formed a similar set of three hydrogen bonds when placed on adjacent strands. These base pairs and their non-covalent interactions (now called Watson-Crick pairs) are what hold the two strands of DNA together in an antiparallel alignment with one another.

- There is even more interaction going on with the base pairs. The pi systems of neighbors stack against each other, creating a strong, contributing dispersion force that holds them together. This interaction has been successfully targeted as a site for antitumor drugs, which can slip in between adjacent bases in rapidly growing cancer cells, causing kinks in their DNA strands, inhibiting their ability to replicate their genetic code and grow.

DNA Fingerprinting

- In recent decades, DNA research has gone well beyond simply characterizing this fascinating structure. Ingenious molecular biologists and chemists have devised new applications using the understanding of DNA. Probably the most well known of these is the art of DNA profiling—identifying an individual and linking him or her to even the smallest hair, skin, or fluid sample, usually for the purpose of forensics.

- There is a common misconception that DNA fingerprinting is statistically irrefutable and that it involves characterizing the entire genome of an individual and comparing it to yet another entire genome from a sample found elsewhere. But this is not the case. The truth is that a DNA fingerprint is a bit less reliable, but only slightly so.

- DNA testing relies on a special protein called a restriction enzyme, which cuts DNA only at very specific sequences of base pairs. The contents of the cut DNA are analyzed by a technique called gel electrophoresis. Essentially, the negatively charged DNA is dragged through a thick gel using an applied voltage. In this environment, larger fragments are impeded more, and smaller ones are less impeded. So, the fragments fan out in a pattern specific to the individual.

Protein Manufacturing

- Another incredible use found for DNA is in the production of proteins for chemical research and as medical therapies. For example, human insulin is a relatively small protein, but it is critical to proper blood sugar management. Adults with type 2 diabetes suffer from an inability to produce the insulin needed to properly manage this critical solute in their blood.

- For decades, sufferers had to inject themselves with solutions of insulin taken from the pancreas of pigs, which is a great solution for most of the population. But a small variation in the pig version of insulin elicited an allergic reaction in a fraction of patients.

- Enter DNA and the world of microbiology. For the past few decades, patients requiring insulin get not the porcine version of the protein, but an exact copy of human insulin— and it does not come from the pancreas of a person, either.

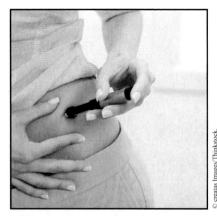

© creatas Images/Thinkstock.

Human insulin is a relatively small protein, but it is critical to proper blood sugar management.

- Scientists take the DNA making up the gene for human insulin and insert it into the DNA sequence of *E. coli* bacteria—a non-pathogenic strain. The bacterial DNA is modified in three important ways. First, genes coding for enzymes to chew up foreign materials are removed. Second, the gene coding for the desired protein is inserted. Third, a special trigger is placed next to that gene. In the presence of the proper signaling molecule, this sequence triggers the bacteria to produce huge quantities of the inserted gene.

- With no protective enzymes to clean up the huge quantity of protein being created, the bacteria are helpless. They express and express and express, until eventually they are not much more than microscopic bags of human insulin. These "bags" are then broken open, and their contents are collected for use in human insulin therapy.

Suggested Reading

McMurry, *Fundamentals of Organic Chemistry*, Chaps. 16.5, 16.6.

Wade, *Organic Chemistry*, Chaps. 23.20–23.24.

Watson, *DNA*.

Questions to Consider

1. Ribonucleotides can mis-pair in DNA when mutations occur. For example, a G-T mismatch can take place when the thymine adopts its enol tautomer. Draw this base pairing between guanine and the thymine tautomer, showing the hydrogen bonds that stabilize it.

2. Before the double helix was proven, what were some of the other proposed structures of DNA? How were they disproven?

DNA and Nucleic Acids
Lecture 23—Transcript

Your physician screens you to determine risk factors for genetic diseases. A criminal forensics lab links a sample of blood, skin, or hair from a crime scene to a suspected criminal. A zoologist or botanist confirms that his or her latest discovery is, in fact, a new species. A pharmaceutical company manufactures human insulin used to treat diabetics using a bacterium as a microscopic production plant.

All of these remarkable accomplishments have one thing in common— DNA. Elegantly simple in its library of just four letters, but stunningly complex in the size and organization of those four letters, DNA and its chemical cousin RNA, stand alone among bio-molecules in their ability to help us carry out these kinds of tasks. That's right, just four letters, that's all it takes to sum up your entire chemical essence. Using just four simple letters, modern scientists can document, store, and even reproduce a code which contains all the basic information which makes you who you are, at least in a physiological sense.

It might sound too simple to be true. Why such a small alphabet of letters to make up a code as complex as the one that makes you, you, in a chemical sense. Well nature makes up with a small library of letters with scale. A single strand of your DNA contains over three billion of these letters. Now, three billion positions with four possibilities for each makes for more possibilities than one can realistically imagine. To get the same number of possible permutations with 26 letters, like our alphabet, would require a literary work equivalent to about 430 copies of Tolstoy's classic, War and Peace.

Most of us are familiar with these four letters, A, G, C, and T. They're a shorthand notation used by geneticists and molecular biologists to represent adenine, guanine, cytosine, and thymine. Now these are the four subunits of a compound which is very familiar to us, DNA. But remember that Voller helped to demonstrate that the chemistry of life is not off limits to the organic chemists. So awesomely complex though cellular biochemistry may be, it is still rooted in the same foundations as the rest of organic chemistry.

The power of understanding and manipulating DNA has been demonstrated time and again since its structure was solved in 1953 by James Watson and Francis Crick. After decades of study, the United States National Institutes of Health and private entrepreneur Craig Venter jointly announced that the human genome had been completely mapped. Now, mapping genes and sequencing DNA has given us a tool which can be used not only to scan for particular health risks in a person, but also to use as a forensic identification tool, a guide for targeted gene therapy, and a way to understand the chemical basis on which life evolves and changes in response to environmental stimuli.

Our understanding of DNA and the biochemical processes surrounding it has altered our perception of what it is to be a living thing. As scientists, we endeavor to study it, understand it, and ultimately, manipulate it in ways which improve our lives. Of course, none of this understanding is possible without tearing this beautiful complex down into its most basic, constituent parts. This is what we will do in this lecture. Starting from the individual molecules and motifs, we will work our way up to a full model for the structure of nucleic acid-containing compounds.

So let's begin with RNA, or ribonucleic acid. RNA serves several critical roles in converting your genetic code, or genome, into the proteins essential to life, also called a proteome. In addition to other roles, one of RNA's most important jobs is to transport the information in the DNA of your cells to areas in those cells where proteins are made. Now, this is a gross underestimation of the versatility of RNA, but for this lecture, we're going to focus on this particular role, since it helps to explain some of the chemical differences between it and its very close chemical relative, DNA.

The ribo part of that name refers to the carbohydrate ribose. Now, the Fischer projection of d-ribose looks like this. It's an aldopentose, and like any aldopentose, ribose can undergo cyclization to form the associated ribofuranose with its anomeric carbon in one of two positions, alpha or beta. It's the beta form of this ribofuranose which is the central building block of RNA. Ribose alone, however, is a fairly uninteresting molecule from a genetic coding perspective. I mean, after all, there's only one beta ribofuranose molecule possible. There's no variation among these structures ... yet.

To produce the building blocks of RNA, the anomeric hydroxyl in beta ribofuranose is replaced with an organic base through a nitrogen atom. In the case of RNA, these four bases are cytosine, uracil, adenine, and guanine. RNA uses uracil as opposed to the thymine of DNA for reasons that we're going to discuss shortly. Now, these units are further subdivided into the class of pyrimidine bases and purine bases based on their backbone structures.

So, using one of these four bases as the aglycone in beta ribofuranose produces one of the four subunits of RNA, collectively called nucleosides. And not surprisingly, these nucleosides have names similar to those of the bases which gave them their identity. An aglycone of cytosine forms the nucleoside cytidine; uracil leads to uridine; adenine produces adenosine; and guanine produces guanosine. So we're slowly constructing the monomers of our miraculous biopolymer. We've attached a puridine or pyrimidine base to ribofuranose as an aglycone to create four different letters in the coding alphabet of life.

But, a simple collection of letters is not a code. Consider this random collection of letters from our alphabet randomly distributed about the screen. There's a little bit of information here in the quantity and identity of the letters, but, I can communicate even more specific and detailed information if I order the letters in a certain way. So let's put the letters together and see if we can figure out what I'm trying to say here.

Teachers get sour, hmmmm, maybe not. Corrugate sheets, well, if this were an engineering course, that would probably be it, but it's not. So let's keep going. The same letters can be put together to form saucers together, which maybe would be a cooking course, I don't know; UFOs possibly; who knows. Or maybe what I'm trying to say is The Great Courses. I can communicate very detailed information with the same letters by combining them in different orders. DNA does the same thing. See, the building blocks of each of these phrases were the same. It's the arrangement of the building blocks which enriches the code and allows me to encode far more information using far fewer sub-units.

So let's go back to RNA. We must now find a way to link those nucleoside sub-units together into a chain so that there's a sequence to the collection of Cs, Us, As, and Gs which will make up the information in our RNA molecule. We need a way to string them together, just like we string letters to make words and sentences. The glue which holds our nucleosides together is a phosphate functional group. Now, I'm going to put a phosphate functional group on these molecules, and you'll notice that the phosphate is sort of a purple color; that's by convention. Now, by attaching a phosphate group at the 5' position on the sugar of our nucleoside, we create a motif called a phosphoester. This phosphoester group will be the linkage through which our nuclesides are connected to give our collection of building blocks order. The collection of three crucial elements, information in the base, the backbone of the ribose linking the glue of the phosphoester all together make up what is called a ribonucleotide. A ribonucleotide is one complete subunit of RNA ready to be linked to others to create a useful code.

So let's put our simple code together consisting of four ribonucleotides condensed into a chain. We'll use our examples of cytidine monophosphate, uridine monophosphate, adenosine monophosphate, and guanine monophosphate. These will be our ribonucleotides. Now, if we align them so that the 5' phosphoester of one can condense with the three prime hydroxyl of another, we can form a phosphoester bridge between the two bases, producing three water molecules in this case as a by-product.

So here I put C, U, A, and G together in a very specific sequence. Now, as a convention, we always read from the 5' end of the chain toward the 3' end of the chain. Of course, this is a very small chain by biochemistry standards, as many RNA strands can be thousands of units long; and they're all performing their functions in your body right now. Among other functions, RNA is nature's messenger molecule; transports information from the genetic code stored in DNA, and it assists in the conversion of that information into the vast library of proteins which our bodies use to regulate their biochemistry.

So, now we understand a bit about the structure of RNA, a complex biomolecule whose job is to act as the middle man, getting information from your permanent genetic code to the site where that code is used to build other compounds. Because the role of RNA is transmission of information, which is carried out on a fairly rapid time scale, RNA itself doesn't need to endure for long periods of time to effectively perform its duties in living systems. In fact, it's part of RNA's design that it can be broken down easily and its components recycled for use in creating more RNA as it's need. RNA already has a pretty stable backbone. But there's a chink in the armor, and it's right here at the 2' hydroxyl group.

This little nucleophile is hanging out right next to the phosphate ester group of the following nucleoside, which has an electrophilic phosphorus atom. So the connection may last for a while, but before long, a 2' hydroxyl of RNA just can't help but attack the phosphorus, initiating a mechanism which ultimately cuts the chain that gives RNA its information-carrying ability. So, RNA can carry huge amounts of information back and forth, being broken down and rebuilt as the organism needs it. But there would be no information to transport or translate if there were no method of storing that information indefinitely until it's called upon. This task of information storage falls to a close chemical cousin of RNA, the famous DNA.

Deoxyribonucleic acid, or DNA for short, is another ribose-based biopolymer which serves a very different purpose. DNA is very similar to RNA, consisting of a phosphate, sugar, and base to form what we call a deoxyribonucleotide. But there are two key structural differences which make DNA more robust than RNA. First, it's the lack of the 2' hydroxyl group of the beta ribofuranose sugar. Instead, this group is replaced with a hydrogen, which is non-nucleophilic. This removal of the one very specific oxygen from the ribose is the source of the deoxyribo portion of DNA's name. With that promiscuous little nucleophile out of the picture, DNA hydrolyzes about 100 times slower than RNA, and this makes DNA a much better choice, not for a short-lived messenger, but for the permanent storage of a genetic code which needs to last you the rest of your life.

A second feature of DNA which makes it a more robust system is the replacement of uracil with thymine, a methylated version of uracil. Now, the reason for this difference is that cytosine itself can deaminate spontaneously to become uracil. Now, obviously, when this happens to your DNA, this is bad news. Your DNA just mutated. But your body has a defense mechanism for this, a complex molecule called an enzyme which finds that uracil in your

DNA and reaminates it to recover the original cytosine. That's amazing, isn't it? Your body can actively repair this kind of damage to DNA.

Now, this incredible repair mechanism comes with a cost. If DNA actually used uracil, how would the enzyme know which uracils to alter and which ones to leave alone? Well it wouldn't. So DNA, instead, uses thymine as the complimentary base for cytosine. The distinguishing methyl group is acting as a signal to the enzyme saying, hey, I'm supposed to be here. Please don't aminate me.

Of course, most of us are familiar with this form of DNA, the double helix. This structure was first proposed by James Watson and Francis Crick, two names which are often associated with it. Of course, the truth is that the discovery of DNA as a double helix was the result of a concerted effort on the part of many talented researchers. In 1869, Friedrich Miescher first isolated DNA from the nuclei of cells. It was this biological source which prompted him to name it nucleon, a name which was later revised to DNA to better reflect its true chemical composition.

Russian researcher Phoebus Levene was able to identify the individual components of DNA, including the sugars, phosphates, and bases which comprised it. But he lacked the necessary tools to properly explain exactly how each of these pieces came together to form a structure for DNA. Moving to the 1940s, Erwin Chargaff developed a method for isolating the bases from the DNA samples. He found that the number of cytosine bases in any given sample closely matched the number of guanine bases. He also determined that the number of adenine bases closely tracked to that of thymine.

In the 1950s, the X-ray crystallography lab of Maurice Wilkins and Rosalind Franklin at King's College finally reached a milestone in genetics. They were able to use a technique called X-ray diffraction to determine the spatial arrangement of atoms within the DNA molecule. In the end, it was Watson and Crick, however, who made it to press first with their model of the double helix, earning them their place in history. What Watson and Crick proposed explained every observation to date. The sugar, the phosphate, and the base molecules found by Levene were all accounted for. The equal amounts of purine and pyrimidine bases from Chargaff's experiments were also easily explained. And the crystallography data from Wilhelm and Franklin's labs, as well as their own, clearly showed that a helical structure held them all together.

The key to putting it all together was that the two pyrimidine bases, adenine and thymine, when held in place by the helix proposed, formed a network of two hydrogen bonds to one another, causing them to pair up. Guanine and cytosine formed a similar set of three hydrogen bonds when placed on adjacent strands themselves. These base pairs and their non-covalent interactions, now called Watson-Crick pairs, are what hold the two strands of DNA together in an anti-parallel alignment with one another, and, that's right, the double helix of DNA is actually held together by nothing more than a few hydrogen bonds. Well, OK, the human genome has about seven or eight billion hydrogen bonds for every strand, so more than a few.

There's even more interaction going on, though, with the base pairs. See, the pi electron system of neighbors stack against each other, creating a strong contributing dispersion force which holds them together even tighter. And this interaction has been successfully targeted as a site for anti-tumor drugs, which can slip in between adjacent bases in rapidly growing cancer cells, causing kinks in their DNA strands and inhibiting the ability of these cells to replicate their genetic code and grow.

DNA is a very exciting structure to study in and of itself, but in recent decades, research has gone well beyond simply characterizing this fascinating structure. Ingenious molecular biologists and chemists have devised new applications using the understanding of DNA given to us by all those who came before. Probably the most well known of these is the art of DNA profiling, identifying an individual and linking them to even the smallest hair, skin, or fluid sample, usually for the purpose of forensics. Now, there's a common misconception that the sheer size of the human genome makes DNA fingerprinting statistically irrefutable, and this is because some believe that it involves characterizing the entire genome of an individual and comparing it to yet another entire genome from a sample found elsewhere. But this is not the case; it can't be, since DNA testing was used long before Venter completed the human genome in 2000. The truth is that a DNA fingerprint is a bit less reliable, but only slightly so.

See, some of the DNA which makes us human is very similar from one person to the next. In particular, the portions of DNA which code for proteins, sometimes called coding DNA, is highly similar, but not all DNA codes for proteins; in fact, much of it is used as a structural support which holds the coding pieces of the sequence in their proper place or performs other non-coding functions. These structural support regions of DNA are sometimes called junk DNA. This is an unfortunate choice of language, because without it, the genes which your body expresses to keep you alive wouldn't function properly. So this hardly sounds like junk to me. Nonetheless, the exact sequence and size of these regions is not as crucial as those comprising coding regions. And nature knows this, so she lets us get away with greater variability in these regions of our DNA.

Now there are several methods for what is commonly referred to as DNA fingerprinting, and all of them focus on non-coding regions of DNA. In all the techniques, a special process called the polymerase chain reaction, or PCR, is used to induce a DNA sample to replicate, creating a huge number of copies of these regions from just a small, original sample. We call this process amplification, and this is why just a vanishingly small physical sample is all that is needed for this kind of technique.

In early methods, a special enzyme, known as a restriction enzyme, was then added to the amplified mixture. The restriction enzyme cuts the amplified DNA but only at very specific points defined by a sequence about a half a dozen or so bases long. This hydrolysis creates a set of fragments whose sizes are unique to an individual. The fragments are then analyzed using a technique known as gel electrophoresis, which uses an electrostatic potential to drag pieces of DNA through a thick gel. This sorts the fragments by size, producing a pattern which is, ostensibly, unique to the individual. More modern testing involves the selected amplification of non-coding regions of DNA from a specific sequence starts, allowing us to bypass the use of a restriction enzyme altogether. But the essential concept remains the same. The statistical likelihood of the number and size of the fragments created by your so-called junk DNA and mine is vanishingly small.

Another game-changing use found for DNA is in the production of proteins for chemical research and as medical therapies. We'll be discussing proteins in detail in our next lecture, but for now I want to focus on just one of them, called human insulin. Human insulin is a relatively small protein, but it is crucial to proper blood sugar management. And adults with type 2 diabetes suffer from an inability to produce the insulin needed to properly manage this critical solute in their blood. For decades, sufferers had to inject themselves with solutions of insulin taken from the pancreas of pigs, which is a great solution for most of the population. But a small variation in the pig version of insulin elicited an allergic reaction in a small fraction of patients.

So it would be much better if we could offer perfect copies of human insulin for therapies like this—enter DNA and the world of microbiology. For the past few decades, patients requiring insulin therapy get not the pig version of the protein, but an exact copy of human insulin. And it doesn't come from the pancreas of a human, either. See, scientists take the DNA sequence which codes for human insulin, and they insert it into the DNA sequence of E coli bacteria. Now, don't panic; this is a non-pathogenic strain of the bacteria, and it's dead long before the therapy is actually conducted on a person.

See, the bacterial DNA is modified in three important ways. First, the native genes in the bacteria which code for enzymes that chew up foreign materials are removed. Second, the gene coding for the desired protein is then inserted. And third, a special trigger is placed right next to that gene that's been inserted. And in the presence of the proper signaling molecule, this trigger will induce the bacteria to express a huge quantity of that inserted gene. With no protective enzymes to clean up the huge quantity of protein being created, the bacteria are utterly helpless. They express and express and express until eventually they're not much more than microscopic bags of human insulin. And these bags are then broken open and their contents collected for use in human insulin therapy. Now, I'll bet you never thought you'd see the day when E coli would be an asset to human health. In truth, it already has been for decades.

The sensational impact that our understanding of DNA has on society makes it a central topic on any lecture covering nucleic acids. But I don't want you to fall into the trap of thinking that DNA and RNA are the only way that these powerful little molecules drive biological processes. A denizen, for example, is not only one of the four nucleotides used to produce RNA, but it's also one of the body's energy storage vehicles. When sugars are oxidized in cells, some of the energy produced is used to produce a denizen triphosphate, which contains three phosphate ester groups attached to the 5' carbon of the nucleotide.

The phosphate ester bonds are somewhat unstable, and that should be immediately evident from the concentration of all that negative charge in such a tiny region. So, breaking just one of them can release a considerable amount of energy, not only as a result of hydrating a phosphate, but also relieving the clash of charges. In the process, about 31 kilojoules per mole, or about 7.3 kilocalories per mole, is released.

Among other things, this ATP to ADP conversion serves as a quick source of stored energy for muscle contraction. This is the process which many athletic supplements claim to target. Most notably, the wildly successful supplement creatine. Now, creatine is used as a vehicle in the body to transport phosphate, which is transferred to ADP by an enzyme called creatine kinase. It's this function that some believe gives creatine supplementation the ability to increase endurance during exercise. Whether creatine supplementation improves physical performance significantly is still a matter of some debate. But what isn't arguable is that it generates hundreds of millions of dollars of annual sales in the U.S. alone, and it's all based on its ability to phosphorylate ADP.

So, let's summarize. In this lecture, we dug deep into the structure of some familiar biopolymers, RNA and DNA. We looked at the structure of ribose and its beta d-furanose form, which makes up part of the backbone of RNA. We saw how four different aglycones attached to this ring lead to the four RNA nucleosides, which we call cytosine and uridine, the pyrimidine-based ones; and adenine and guanine, the purine-based nucleosides.

Then we took a look at the phosphate ester linkage, which starts with the 5' carbon of a nucleoside, making it a nucleotide. When connected to another nucleotide through its 3' hydroxyl, we have the makings of RNA, the molecule responsible for translating the genetic code into proteins. After that, we considered the source of that genetic code, DNA. We saw how DNA is tailored to provide a long-lasting informational template by the removal of the 2' hydroxyl group from ribose to slow hydrolysis and by replacement of uracil with thymine to provide a system which can be easily repaired when cytosine is damaged by deamination.

We saw how the double helix was discovered and how it explains so much about the previous 90 years of observations by geneticists and chemists. Next, we looked at how DNA can be used as a substitute for fingerprinting, providing a much more easily left and much more accurately matched identifier for victims and perpetrators of crimes. We also saw how genetics has given us a way to use DNA to manipulate the biochemistry of bacteria, exploiting their biochemistry to create therapeutic agents for humans.

And finally, we looked at how nucleotides themselves can serve as energy storage units in living systems. Specifically, we saw how the conversion from ATP to ADP releases a large amount of energy, which the body can use for muscle contraction, and how creatine and creatine phosphate can help to reload this reaction so it can be used again. This prompts athletes around the world to munch down grams of this organic compound on a daily basis.

In our next lecture, we will concern ourselves with the downstream end of biology's central dogma—proteins. We will learn about the structure of amino acids and how they come together, both in biological systems and in the lab, to create the powerful enzyme catalysts that I've described to you in this lecture. I'll see you then.

Amino Acids, Peptides, and Proteins
Lecture 24

I n this lecture, you will learn about a class of materials that not only make up 20% of the mass of your body, but that find use in medicine, materials, drug chemistry, detergents, and many more functions. This class of materials is proteins. Usually associated with skin, hair, and muscle fibers, proteins make up a class of catalysts that mediate every chemical reaction that your body carries out. But before you can develop an appreciation for these large, complex kings of catalysis, you have to become familiar with their building blocks—amino acids.

Amino Acids

- Amino acids are so named because this class of compound always has two distinct functional groups: an amine and an organic acid joined by one or more carbons. The simplest possible example is an amino acid in which amine and acid groups are joined by a single CH_2 (methylene) group. When only one carbon separates the functional groups, we call these "alpha" amino acids.

- Extending the chain by one more methylene creates a "beta" amino acid, and yet another insertion creates a "gamma" amino acid, and so on. Alpha amino acids are the ones that were chosen by nature as the basis for proteins; they are a very special set of compounds that carry out the chemistry of life.

- The simple alpha amino acid glycine has two different methylene hydrogens on its alpha carbon. It is this position that is modified to create a short library of 20 alpha amino acid compounds used to produce nearly all of the proteins and enzymes that drive the chemistry of life. When something other than a hydrogen is present in one of these positions, we refer to the group as a side chain.

- Of course, there is really no side, front, back, top, or bottom to a molecule, but when we align various amino acids for the purposes of comparison, we usually line up the amino acid portion of the molecule and treat the side chain as a motif branching off to the side. Even so, these bonds can twist and rotate and do not necessarily exist locked in a specific configuration.

- Side chains can be broken down into several classes based on their chemistry. For example, aliphatic (or hydrocarbon) side chains are present in the amino acids alanine, valine, leucine, and isoleucine. Acidic side chains are available on aspartic acid and glutamic acid. Hydroxyl groups make an appearance in threonine and serine. Nitrogen-containing side chains are on lysine as well as arginine and histidine.

- Aromatic side chains include phenylalanine, tyrosine, and tryptophan. Asparagine and glutamine have terminal amides. Methionine and cysteine contain sulfur atoms. Finally, proline is a curious amino acid whose side chain wraps back around and bonds to its own amine nitrogen.

- All but glycine contain a chiral center at the alpha carbon, meaning that there are two possible enantiomers of each compound. Remarkably, all 19 amino acids with a chiral alpha carbon take on the same stereochemistry in living systems. So, only one enantiomer of each amino acid is useful biologically.

- The reason for this is twofold. First, chiral RNA is responsible for sequestering and using these amino acids to build proteins, so only one handed form interacts properly with the ribosomal RNA. The other reason is that many amino acids are manufactured from one another in the body. Only 8 of the 20 are absolutely essential in your diet. The rest can be made by chemically modifying some of the 8 essentials.

Synthesis of Amino Acids

- There are quite a few ways to produce amino acids from non-biological sources, though biological sources are very attractive because amino acids make up a very large percentage of the biomass of living systems.

- In 1850, German chemist Adolph Strecker devised a way to produce amino acids from aldehydes using ammonium chloride and potassium cyanide. His process generates an amino nitrile via an iminium intermediate. Ammonia is then eliminated with the use of an aqueous acid.

- The most noteworthy amino acid synthesis was the one carried out by young graduate student Stanley Miller in the 1950s at the University of Chicago. Miller was a student of Harold Urey, who had a keen interest in the origin of life and had asked if it were possible for simple, vital organic compounds to form under the reducing, stormy conditions of the Earth's early atmosphere.

- Miller designed a system that simulated the early Earth's atmosphere in closed conditions on a lab bench. He enclosed water, methane, ammonia, hydrogen, and formaldehyde in a sealed vessel and allowed an electrical current to arc through the mixture for an extended period of time.

- After running his apparatus for a week, Miller tapped the circulating solution and analyzed it. When he did, he found a broad range of amino acids. Miller himself detected 11 of the 20 amino acids necessary to support life as we know it. In later years, sealed samples from his experiments have been analyzed and shown to contain all 20.

Peptides and the Amide Bond

- When we put amino acids together via a series of condensation reactions that form chains of amide bonds, the polyamides that result are called either peptides or proteins. The distinction between the two is one of size, with the former generally referring to polymers of less than 100 residues and the latter encompassing larger molecules.

- For example, if we could condense two glycine residues together in an end-to-end relationship, we would create a very short chain consisting of two glycines, minus a water. The loss of the water is important, because it balances what would be an entropically disfavored reaction.

- We can continue the process, condensing another to make triglycine, then another, and so on. The result is a chain of glycine residues. We call them residues, instead of amino acids, because they have lost one equivalent of water when they are in the chain.

- Biochemists like to arrange their polyamides so that the nitrogen-containing terminus is at the left, and the sequence can be read like a sentence from left to right—N terminus to C terminus.

- Our polyglycine peptide is pretty boring as peptides go, but nature has provided us with 19 more options for each slot in the peptide. It is this variety that leads to stunning complexity in the possible sequences for peptides. We call each unique sequence of condensed amino acids a primary structure.

- Furthermore, when amino acids condense to form an amide bond, the amine nitrogen from one residue and the carbonyl from the preceding residue participate in a resonance that locks in a plane defined by the N, H, C, and O of the amide bond. We call this collection of atoms the amide plane.

- But that leaves the alpha-carbon bonds free to rotate, and this gives proteins a tendency to twist and collapse on themselves in very specific arrangements that maximize intramolecular attractions created by backbone atoms and side chains as well. Not only do alpha carbon backbone bonds rotate, but they tend to do so in just a few specific geometries, called secondary structures. These structures include many different versions of three basic motifs: helixes, strands, and turns.

The Folding Process

- Higher-order elements of structure—secondary, tertiary, quaternary—are in theory dictated simply by the sequence, or primary structure, of a protein. If we simply construct the linear protein for nearly any naturally occurring enzyme, we often find that the protein will most likely fold into its functional form—usually with startling speed.

- In rare instances, however, the functional form of the protein is actually not the most stable fold. Proteins can become misfolded and trapped in a topology that will not allow them to fold properly. Ordinarily, it wouldn't be too alarming if a single protein in your body misfolded. After all, you have millions and millions more to get the job done.

The misfolding of proteins is responsible for the buildup of plaque in the brains of Alzheimer's victims.

- But it does become disconcerting when one misfolded protein encourages the misfolding of others, producing plaques of proteins with potentially devastating health implications. This situation most notably happens when amyloid plaque builds in the brain of an Alzheimer's disease victim.

Protein Synthesis

- Biologically synthesized proteins can be produced using complex RNA machinery within cells, but what do chemists do when they want to make highly purified peptides and proteins in the laboratory for study? Several techniques are available, including the use of special strains of *E. coli* bacteria as manufacturing plants for proteins.

- The genome of a very special, already genetically modified *E. coli* is altered to encourage the bacteria to produce massive quantities of just one desired protein. This technique is startlingly effective when large proteins containing naturally occurring amino acids are the target.

- There is a method for synthesizing even peptides and proteins with exotic side chains using organic chemistry. The developer of this method, Bruce Merrifield, won the 1984 Nobel Prize in Chemistry for his development, which is called solid phase peptide synthesis. It is sometimes also called the Merrifield synthesis.

- Merrifield's technique starts with an insoluble polymer resin to which a single amino acid is attached. The connection to resin is made through the C terminus, and the N terminus is available for reaction. Merrifield then turned to a classic technique for making amide bonds—an active ester coupling. By mixing amino acids with a special reagent that turns them into esters, the exposed amine of the growing chain can be enticed to attack the carbonyl of amino acids in solution.

- This is a problem when the amino acids in solution have amine groups of their own. If unchecked, the growing peptides will extend in an uncontrolled fashion. So, Merrifield applied a special protecting group to the N terminus of the amino acids prior to reacting them with the growing peptide on resin. Merrifield used a group called a *t*-butoxycarbonyl motif.

- When present, this N-terminal protecting group prevents more than one amino acid from attaching to the growing peptide in a single reaction. The excess reagent can then be washed away from the resin, and the N-terminal protecting groups can be removed before introducing the next coupling. In this way, Merrifield was able to produce peptides in sufficient yield and purity to be used in experimentation.

Suggested Reading

McMurry, *Fundamentals of Organic Chemistry*, Chaps. 15.3, 15.4.

Wade, *Organic Chemistry*, Chaps. 24.8–24.13.

Questions to Consider

1. Although some peptides and proteins hydrolyze easily under acidic conditions, some are remarkably stable. Why?

2. Some enzymes become inactive or insoluble in water at different pH values. Why?

Amino Acids, Peptides, and Proteins
Lecture 24—Transcript

So, we've seen how Friedrich Woehler's synthesis of urea from non-biological starting materials in 1828 toppled the long-standing dogma known as vitalism, which stated that only organisms could produce the molecules of life. Now, as the 1800s pressed on, other notable chemists advanced this idea by putting it into practice. Justus von Liebig demonstrated the role of abiological nitrogen in plant fertilization, and he also extracted and manipulated carbohydrates under controlled laboratory conditions. Miescher had isolated DNA from the nuclei of cells. And as the century drew to a close, Alexander Butlerof and Herman Emil Fischer both managed to synthesize carbohydrates and other biologically relevant compounds in their laboratories, also without the help of living organisms.

So, by the end of the 1800s, organic chemists were beginning to get a hold on exactly how to produce, characterize, and even manipulate the chemicals of life abiologically. One could argue that today we have successfully, if not yet completely, dissected life all the way down to the molecular level. We can take molecules out of living systems and use them in vitro. We can synthesize compounds like carbohydrates, RNA, and DNA in labs and insert them back into living systems to show that they function exactly as one would predict for their biologically synthesized equivalents.

But there remains one great burning question which is unanswered to this day—If all of these molecules of life are truly devoid of any vital essence, shouldn't we be able to use those synthetic compounds not just to sustain or manipulate existing life, but to create life itself? Now, life on Earth must have originated by some accidental, abiological synthesis of some of the compounds necessary to achieve life. But what were those compounds, and how did they come to be with no living organism or human scientist around to create them? It's an intriguing question and one which is difficult to answer, considering that it took place at least 2.7 billion years ago, and nobody was around to see it happen. Was it DNA? RNA? Maybe carbohydrates or amino acids. And, whatever these seminal biomolecules were, how did they come together in a way which allowed them to cross that critical barrier of simply existing to self-replication, allowing them to reproduce, grow, and eventually evolve?

So scientists started on a new quest, not one to learn how biomolecules function per se, but rather, how they came to be in the first place. And two of these researchers were Harold Urey and Stanley Miller. Miller was a student of Urey in the 1950s. Now, already a Nobel laureate at the time, Urey had developed an interest in the origin of life and had asked if it were possible for simple, vital organic compounds to form under the reducing, stormy conditions of the Earth's early atmosphere.

Miller had just arrived at the University of Chicago as a graduate student and was slated to work on a project aimed at understanding the synthesis of elements in the furnaces of stars. But Miller found himself inspired by Urey's interest in the origin of life and asked for permission to try a different experiment, one to imitate the conditions of the primordial Earth and probe the resulting material for signs of biologically relevant molecules.

Urey agreed, and Miller went straight to work designing a system which simulated the early Earth's atmosphere in closed conditions on a lab bench. The conditions on the Earth some three billion years ago were very different from those today. Volcanism was rampant, oxygen was scarce, and electrical storms were commonplace. So to model this, Miller devised an apparatus which enclosed water, methane, ammonia, and hydrogen in a sealed vessel. Now, he allowed an electrical current to arc through that mixture for an extended period of time as the water was boiled. After running his apparatus for a week, Miller tapped the circulating solution and analyzed it, and sure enough, when he did, he found evidence of several amino acids.

Now, Miller published the results in the prestigious journal *Science* in 1953, under the title "A Production of Amino Acids Under Possible Primitive Earth Conditions." And remarkably, the paper is published citing Miller as the lone author. This was done at Urey's insistence, lest he receive the credit for Miller's ingenious experiment.

So perhaps it was amino acids which were the very first biological molecules to participate in life on Earth. But if they were, these are some pretty small molecules, and life would need something larger, more complex, to really generate diversity and get a foothold. But amino acids are able to create these kinds of molecules by joining together into long strands through a special linkage known as an amide bond.

During this lecture, we're going to talk about a class of materials which not only make up 20 percent of the mass of your body, but which find use in medicine, materials, detergents, and many other functions. Today we're going to talk about proteins. Usually associated with skin, hair and muscle fibers, proteins are really much more than that. They also make up a class of catalysts which mediate every chemical reaction that your body carries out. But before we can develop an appreciation for these large, complex kings of catalysis, we have to review their building blocks, amino acids.

Amino acids are so named because this class of compound always has two distinct functional groups, an amine and an organic acid, joined together by one or more carbons. Now, the simplest possible example is an amino acid in which the amine and acid groups are joined by a single CH_2 group, also called a methylene group.

When only one carbon separates the functional groups, we call these alpha amino acids, and you can imagine that extending the chain by one more methylene creates a beta amino acid, and yet another insertion creates a amine amino acid, and so on. Of course, I've drawn these compounds as completely neutral species, but each contains a carboxylic acid with a pK_a of about 5 and an amine whose conjugate base has a pK_a of about 11.

So a more proper way to draw these compounds might be like this, with an intramolecular proton transfer reaction creating two regions of charge. Now we call these species zwitterions. The charged regions of these species don't interact favorably with most low-polarity organic solvents. So, they can make the solution in organic solvents a painful process for amino acids. It's a good thing that you're made mostly of the very polar solvent water, so they dissolve in you quite readily. Although there are many other ways to connect an amine group and a carboxylic acid, in today's lecture we will keep our attention squarely focused on alpha amino acids, because this class of amino acids was chosen by nature as the basis for proteins, a very special set of compounds which carry out the chemistry of life.

So, getting back to our simple alpha amino acid. We commonly call this little guy glycine. It's one of the first amino acids detected by Stanley Miller. Now, notice that glycine has two hydrogens on the alpha carbon. It's this position which is modified to create a short library of 20 alpha amino acid compounds, which are used to produce nearly all of the proteins and enzymes that drive the chemistry of life. When something other than a hydrogen is present in one of those two positions, we refer to the group replacing that hydrogen as a side chain. Now, there's really no side, front, back, top or bottom to a molecule, of course, but, when we align various amino acids for the purpose of comparison, we usually align the amino acid portions of the molecules up and treat the side chain as a motif branching off in the same direction, hence the name.

Even so, we know that these bonds can twist and rotate and do not necessarily exist locked in the configurations that I'm going to show you. Side chains can be broken down into several classes based upon their stereochemistry. OK, let's take a look at a few of these amino acids now. We've already discussed a little bit about the simplest of all amino acids, which would be glycine. Glycine is unusual among the amino acids, because it has two of the same groups attached to its central carbon, or its alpha carbon. So in this case, glycine is unusual because it's really unmodified.

But as we began modifying one of those groups, we start to create our library. Adding a methyl group creates alanine, another one of the amino acids which was detected by Stanley Miller. If I continue to increase the complexity of that side chain, I can create valine, leucine, and isoleucine. As you can see here, I've highlighted all of their side chains by sort of a grayish glow. The reason I've grouped these five together is that their side chains are hydrophobic. In other words, they're all made strictly of carbon and hydrogen, and so they behave as though they were very low polarity arms on the molecule. But that's only true for these five. Let's take a look at some others.

Aspartic acid and glutamic acid are acidic side chains, meaning that in most physiological environments you actually see them as their conjugate bases, so they have a tendency to carry negative charges. Asparagine and glutamine are two amino acids which have amide functional groups at the ends of their chains, so these are specialists at hydrogen bonding to other side chains and other molecules. The amino acids lysine, histidine, and arginine all contain slightly basic amine groups, or in this case, histidine has a cyclic side chain. Nonetheless, it has a nitrogen, which is basic. And so all three of these can become fairly easily protonated and therefore carry a positive charge. So these are the basic side chains of the amino acid library.

Continuing on, we have serine, threonine, cystine, and methionine, all of which contain either an OH, an SH group known as a thiol, or the group known as a thioether, where the sulfur is contained in between two carbons. The reason I placed all of these together is that they all contain fairly good nuclea files, so, these four are particularly good at creating covalent bonds between other molecules and their own side chains.

Moving on, we have aromatic side chains as well in the form of phenylalanine, tyrosine, and tryptophan. All of these amino acids contain large, conjugated, or more accurately, aromatic, phi systems, which can be used to create all sorts of different interactions that you wouldn't get otherwise using, say, hydrophobics, or any other amino acid for that matter.

And finally we have proline, which stands alone among amino acids because of a very special feature. Proline's side chain actually wraps around and bonds back to the nitrogen of the amino acid. So, proline is unusual in the sense that its side chain is locked in a cyclic confirmation, rather than being free to move attached to an sp^3 carbon. So proline is going to have some very special roles in biochemistry.

But we're still not done characterizing these amino acids. Now, you probably noticed that all but glycine contained a chiral center at the alpha carbon, meaning that there are two possible enamtiomers of each compound. Now, what's remarkable is that all 19 amino acids with a chiral alpha carbon take on the same stereochemistry in living systems. So only one enantiomer of each amino acid is useful biologically. The reason for this is two-fold.

First, chiral RNA is responsible for sequestering and using these amino acids to build proteins. So, only one handed form interacts properly with the ribosomal RNA.

The other reason is that many amino acids are manufactured from one another in the body. In fact, only 8 of the 20 that I just showed you are absolutely essential in your diet. All the others can be made in your body from other starting materials, sometimes by chemically modifying some of the 8 essential amino acids.

So, why was Miller's discovery so sensational? What is it about amino acids which make their presence in his primordial concoction so newsworthy? Well, now that we have a firm grasp on the structure of amino acids, let's consider what happens when we start putting them together via a series of condensation reactions which form chains of amide bonds. We call these polyamides peptides or proteins. And the distinction between the two is one of size, with the former generally referring to polymers of less than a hundred condensed amino acids, and the latter encompassing larger molecules.

So let's begin with just glycine. If I could condense two glycine amino acids together in an end to end relationship, I would find that I have created a very short chain consisting of two glycines, minus a water molecule, of course. Now, I can continue the process, condensing another, to make a triglycine, and another, and so on, for as long as I like. The result will be a chain of glycine residues. And we call them residues instead of amino acids since they've lost one equivalent of water when they're placed into the chain. Biochemists like to arrange their polyamides so that the nitrogen-containing terminus is at the left and the sequence can be read like a sentence from left to right, from end terminus to C terminus.

Now, our polyglycine peptide is pretty boring as peptides go. But remember that nature has provided us with 19 more options for each position in that peptide. It's this variety which leads to the stunning complexity in the possible number of sequences for peptides. Let's consider a protein of 100 residues, fairly short by biological standards. This sequence has 100 positions, with 20 possibilities in each position, for 20^{100} number of possibilities; that's roughly

10^{130}, far more possibilities than there are atoms in the known universe. So, clearly, variety is not a problem when it comes to proteins and peptides.

Now, we call each unique sequence of condensed amino acids a primary structure. But the rabbit hole goes much, much deeper than that. You see, when amino acids condense to form an amide bond, the amine nitrogen from one residue and the carbonyl from the preceding residue participate in a resonance which locks a plane defined by the N, HC, and O of the amide bond. We call this small collection of atoms the amide plane. But that leaves the alpha carbon bonds free to rotate, and this gives proteins a tendency to twist and collapse on themselves in very specific arrangements which maximize intramolecular attractions created by backbone atoms and those side chains as well. We call this process protein folding. Now, not only do alpha carbon backbone bonds and side chain bonds rotate, but they tend to do so in ways which form just a few specific geometries, called secondary structures. These structures include many different versions of essentially three basic motifs—helices, strands, and turns.

So, let's take a look at the most common versions of each of these now. We'll start with a short polyaniline molecule. Polyanilline tends to form a secondary structure known as an alpha helix. In this example, I've created an idealized helix, which is perfect in all angles and interactions. The alpha helix is characterized by a helical pitch of 3.6 residues, meaning that one full turn around the helix is completed for every 3.6 residues that I move along the chain. What this pitch accomplishes is to bring the amide NH of one residue into proper alignment to hydrogen bond with the carbonyl four residues down the chain.

Now, we can see that the alpha helix is in the structure of a myoglobin. If we include all of the atoms in a molecule of myoglobin, it's very difficult to see. But if instead we remove side chains while leaving their effects on the structure in place, we can begin to see the repeating turns of quite a few helices in the protein. But, biochemists, not to be outdone, like to make things prettier than organic chemists, so they like to show these colorful 3-D ribbons, which map out the backbone of a peptide or protein, making it look more like this. Now the secondary structural elements are unmistakable, and I have to admit, beautiful.

Another common structure is the beta strand, in which the peptide chain is fully extended. Now, this causes the chain to take on a very long, extended geometry, which may seem a bit implausible at first, but the driving force for this substructure becomes clear when we align two or more of them anti-parallel to one another, creating a network of NH to carbonyl hydrogen bonds in between the strands. The resulting network is referred to as a beta sheet and appears in many well-characterized proteins.

A third type of motif, and incidentally, one which is all but essential to make a beta sheet, is a turn. There are many phi and psi angles, that is, the angles among those atoms in those rotating alpha bonds, alpha carbon bonds, which can accomplish turns in proteins. And certain amino acid residues seem better suited to producing them. Among others, glycine and proline are very common turn residues for very different reasons. Glycine, because the lack of a side chain allows it a greater range of phi and psi angles, meaning more flexibility to turn; and proline, because this cyclic nature of the chain restricts it to just a few phi and psi angles, which are consistent with turns. So it has the tendency to disrupt the patterns in chains and sheets so that a turn can begin.

But the structural complexity doesn't end there. Notice that each element of secondary structure packs against other secondary structures in a folded protein. The interactions between the faces of these two elements of secondary structure is a stabilizing factor as well and brings different portions of the sequence in the proper proximity and orientation for proper protein function. We call the arrangement of secondary structural elements in space a protein's tertiary structure.

Additionally, entire proteins can act as sub-units to a single, large complex, such as this voltage-gated potassium ion channel characterized by Roderick MacKinnon in 1998. This conduit, designed to span a cell membrane and create a channel through which ions can flow, only functions when all eight of its protein sub-units are mated together through non-covalent interactions, creating a super structure, which forms a tunnel traversing the membrane of a cell. This tunnel is lined with the oxygen atoms of amide bonds from the protein's backbone. This creates a negatively charged conduit, which is exactly the width of a potassium ion. And these very specific primary, secondary, tertiary, and quaternary structures all combine to form a molecular device with powerful selectivity for potassium ion transport.

So, all of these higher-order elements of structure—quaternary, tertiary and secondary—are in theory dictated simply by the sequence, or the primary structure, of a protein. If we simply construct an unfolded protein for nearly any naturally occurring enzyme, we often find that protein will most likely fold right back into its functional form, usually with startling speed. In rare instances, however, the functional form of the protein is actually not the most stable fold. Proteins can become misfolded and trapped in an arrangement which will not allow them to fold properly.

Now, ordinarily it wouldn't be too alarming if a single protein in your body were to misfold. After all, you have millions and millions more to help get the job done. But it does get disconcerting when one misfolded protein encourages the misfolding of others, producing plaques of proteins which potentially can have devastating health effects. And this situation most notably happens when amyloid plaque builds in the brain of an Alzheimer's disease victim.

OK, so we clearly can benefit from understanding how protein sequence affects protein folding. The next order of business is to figure out how to make samples for study. Of course, biologically synthesized proteins can be produced using complex RNA machinery within cells, but, what do chemists do when they want to make highly purified peptides and proteins in the laboratory for study? Well several techniques are available, including the use of spatial strains of E coli bacteria as manufacturing plants for proteins, as we saw in the manufacture of human insulin. We saw in our DNA lecture how that genome of that very special, already genetically modified E coli was altered to encourage the bacteria to produce massive quantities of that one desired protein.

Now, this technique is startlingly effective when large proteins containing naturally occurring amino acids are the target. But this isn't a microbiology course; it's a chemistry course. So, there's a method for synthesizing even peptides and proteins with exotic side chains and very interesting organic chemistry. Absolutely. And its developer, Bruce Merrifield, won the 1984

Nobel Prize in chemistry for his development called "Solid Phase Peptide Synthesis," sometimes also called the Merrifield synthesis.

Merrifield's technique starts with an insoluble polymer resin to which is attached a single amino acid. The connection to the resin is made through the C terminus, and the N terminus is available for reaction. Now, Merrifield field then turned to a classic technique for making amide bonds, called an active ester coupling. By mixing amino acids with a special reagent which turns them into very electrophilic esters, the exposed amine of the growing chain can be enticed to attack the carbonyl of an amino acid in solution. Of course, this is a problem when the amino acids in solution have amine groups of their own. If unchecked, the growing peptides will extend in an uncontrolled fashion. So Merrifield applied a special protecting group to the N terminus of the amino acids prior to reacting them with the growing peptide on the resin.

Now, Merrifield used a group called a t butoxycarbonyl motif. When attached to an amide, this N terminal protecting group prevents more than one amino acid from attaching to any given growing peptide in a single reaction step. The excess reagent could then be washed away from the resin and the N terminal protecting groups removed from the on-resin peptide before introducing the next coupling reagent. In this way, Merrifield was able to produce peptides in sufficient yield and purity to be used in lab experimentation.

So let's review what we've covered. We've discussed amino acids, most extensively, alpha amino acids. We looked at the structural features which define them, and we identified the side chains of the 20 amino acids which are necessary to maintain life. We talked about the Miller experiment and how it can be used to create amino acids abiologically, though at best it can produce racemic mixtures or achiral amino acids.

We also covered how amino acids can be condensed together into chains of varying length joined by amide bonds. And we defined the amide plane and the backbone dihedral angles phi and psi associated with rotation of the carbon-carbon bond at the alpha carbon. We looked at how different phi and psi angles produce backbones leading to different secondary structures.

Specifically, we looked at alpha helices and beta sheets and the network of hydrogen bonds which hold them together.

We also talked about turns, and how proline and glycine tend to concentrate themselves in those areas. We talked about tertiary and quaternary structures in proteins as well, and we looked at very large complexes consisting of multiple protein sub units, like the one characterized by Rod MacKennon.

Finally, we addressed the issue of creating proteins in the lab so that we can continue to study the structure and interaction of these compounds on a lab bench, with the hope that we can someday relate what we have learned to how they behave in organisms.

So, we've finished our quick, three-lecture survey of large biological molecules, most of which can simply be thought of as very large organic compounds consisting mostly of the same elements as the smaller molecules making up those classic small molecules of our science.

The time has come for us to consider some new additions to organic chemistry—metals. In our next lecture, we'll move slightly away from organic chemistry's interface with molecular biology and focus more on its connections with the chemistry of metals. We'll tackle this special class of compounds called organometallics next time. And I'll see you then.

Metals in Organic Chemistry
Lecture 25

I t is just about impossible to deprotonate an alkane under any kind of sensible laboratory conditions. But there is a class of compounds that can be made to act like carbanions. In order to do this, we have to abandon the safety and comfort of the first two rows of the periodic table. We have to use a remarkable marriage between the usual suspects of organic chemistry and metals. In this lecture, you will investigate a small slice of a field known as organometallic chemistry.

The Birth of Organometallic Chemistry

- Organometallic chemistry got its start in the lab of Aleksandr Butlerov at the University of Kazan in Russia in the 1800s. Butlerov had noticed that dialkyl zinc reagents, chemicals containing two organic alkyl groups bonded directly to a zinc atom through their carbons, could be used to produce alcohols from ketones by transferring their alkyl group to the carbonyl carbon of the ketone.

- Excited about the prospect of developing a new method for the formation of carbon-carbon bonds, Butlerov passed this project on to his protégé, Aleksandr Zaitsev, who was able to apply its chemistry to several new substrates, but always with limited success. These zinc-based reagents proved to be unpredictable and difficult to work with.

- Then, in 1900, a young French graduate student by the name of Victor Grignard blew the lid off of organometallic chemistry by devising a powerful, versatile, and adaptable organometallic reagent for use in producing new carbon-carbon bonds.

- In the late 1890s, in Philippe Barbier's lab at the University de Lyons in France, while Grignard was conducting experiments for his doctoral dissertation, he acted on a hunch. He replaced the zinc from the only known class of organometallics of the day with what he believed would be more reactive magnesium. His work not only earned him his doctorate, but one-half of the Nobel Prize 12 years later. His corecipient was Paul Sabatier, for his metal-catalyzed hydrogen-addition reaction.

The Grignard Reaction

- So, what had Grignard accomplished? He mixed an alkyl halide with magnesium metal in ether. This combination of reagents produced what we would call an alkyl magnesium halide, in this case an alkylmagnesium bromide. This process involves two distinct radical reactions that take place when the alkyl halide bond spontaneously and breaks, sending one electron each way.

- So, we have two radicals, hungry for a partner for that unpaired electron. We also have neutral magnesium metal, with two valence electrons that it is more than willing to give up to establish its own octet. The magnesium strikes a deal with the two radicals and inserts itself between the alkyl group and the halide, forming two new covalent bonds—one to the halide radical, and then one to the alkyl radical. From an octet perspective, everyone is happy again.

- But in regard to the covalent bond between the carbon and magnesium, magnesium has a very low electronegativity—so low, in fact, that we usually refer to it instead as having high electropositivity. Its electronegativity being so much lower even than carbon, the new carbon-magnesium bond has a dipole oriented with the negative end of the dipole on the carbon.

- It is this increased negative charge density that made Grignard reagents act much like analogous aliphatic carbanions, but without the hassle of using some of the strongest bases to get it.

- If we mixed isoamyl anion with carbon dioxide, what would we get? How about a nucleophilic attack on the carbon of CO_2, followed by a quick workup in acid to protonate the carboxylate we just formed? If we do this, we get isocaproic acid, the exact compound that Grignard had produced using isoamyl bromide, magnesium, and CO_2.

- Of course, when Grignard made it, his reagent was not a carbanion, but an alkyl magnesium bromide reagent, which means that we have to account for the magnesium and bromine. What happens to them during the nucleophilic attack is a bit complicated and case-specific, but the simplest way to model the reaction is with the Mg-Br motif attached to the oxygen of the new carboxylate or alkoxide. Exposing these species to aqueous acid leads to carboxylic acid and magnesium hydroxybromide.

- Grignard's new nucleophile could be used to produce acids from carbon dioxide, secondary alcohols from aldehydes, and tertiary alcohols from ketones. It seems that anything with a carbonyl is fair game for them. Grignard reagents are extremely versatile, offering not only ways to make alcohols and acids, but also aldehydes from amides, tertiary alcohols from esters, and many more.

- But there is much more to the Grignard reaction. Easy though it is to run, there are a few special considerations that must be taken to ensure its success. First, the Grignard reagent itself is only stable in ether solvents like diethyl ether or tetrahydrofuran. Second, the reaction mixture must be kept extraordinarily free of protic solvents, including water.

Ethers and Protic Solvents in Grignard Syntheses
- One of the great limitations of Grignard reactions is that alkylmagnesium halide reagents can only form in ether solvents. But why? The answer lies in the ability of solvents to coordinate metals. The solvent must have oxygen atoms with lone pairs that can stabilize the Grignard by donating electrons to form a stabilizing metal-oxygen interaction in solution.

- That alkyl magnesium halide motif is so polar that it needs a solvent that will be able to arrange its dipole to beneficially interact with the positive charge density at the magnesium. For this purpose, ethereal solvents are just the ticket. Nonpolar solvents like hexanes or benzene are not capable of forming such stabilizing interactions, so they are clearly out.

- Grignard reagents are really polar, so why can't we use more polar solvents like ethanol to dissolve them for reactions? Ethanol has enough polar punch to get this reagent to form in solution, but it is protic, with a pK_a of about 17. Recall that Grignard reagents react like carbanions, whose conjugate acid has a pK_a of about 60. So, a Grignard reagent will form in ethanol—for a split second before reacting with the ethanol to form ethoxide and a hydrocarbon. This is not a good way to make a Grignard.

- Ethers like diethyl ether and THF are special because they strike a balance between their ability to solvate a Grignard with their inability to react with it, and this is why they have been the preferred solvent for this sort of reaction from 1900 until today.

- But using ether and THF isn't enough. Because it is slightly polar, ether can dissolve a small bit of water. Our solution to this is to use ether that has been chemically dried before use. We accomplish this through the use of inorganic drying agents like magnesium sulfate or molecular sieves, which act like sponges, absorbing the water from the ether and trapping it as a solid complex that can be filtered or decanted away.

- A more modern substitution for many Grignard reagents is *n*-butyllithium, which itself can by made by mixing *n*-butylchloride or bromide with lithium metal and is preferred over other alkyllithium reagents for its relative stability. Bottles of this reagent can be purchased and stored for short periods of time, while Grignard reagents usually must be prepared at the time of use.

Organometallic Catalysts

- In addition to the use of metals to activate hydrocarbons and turn them into nucleophiles, another use of organometallic compounds harkens back to Paul Sabatier, who developed a technique for hydrogenating fats using platinum metal.

- In modern laboratories, the reduction of alkenes to alkanes is still a widely used technique, but the heterogeneous nature of Sabatier's methods leaves much to be desired. Hydrogenating double bonds at the surface of a solid catalyst may work well on industrial scales, but in the laboratory, a reaction that takes place in solution is more desirable.

- If we had access to a homogeneous catalyst—one that could be dissolved in organic solvents along with the substrate—we could run reactions much more efficiently and cleanly in the lab. But the problem is that metals themselves are not generally soluble in the kinds of solvents that one would use to dissolve an alkene.

- In 1966, British chemist Geoffrey Wilkinson published a paper outlining a way to solubilize a large transition metal like rhodium by bonding to it organic molecules that we call ligands. These ligands were specially designed to be strong Lewis acids, which would donate stabilizing electron pairs to the metal ion, forming a complex that gave it the solubility characteristics of the ligands while regaining the catalytic activity of the metal.

- Wilkinson's work is generally regarded as the seminal work in the field of organometallic chemistry. His discoveries in this field launched an age of investigation into metal-ligand interactions that continues to this day in labs around the world.

- Wilkinson's work has inspired the creation of some really exciting and exotic-looking complexes. You will see extensions of his work in such life-changing compounds as contrast reagents for MRI.

Biological Organometallics

- Some of the most astonishingly powerful organometallic catalysts are the ones at work in your body right now. For example, if we remove all of the protein atoms from a subunit of the structure of human hemoglobin, left over is a large, cyclic ligand coordinating an iron atom from four planar positions. It is the coordination of these four nitrogen atoms, and also two others in the form of histidine amino acid residues, that gives the molecule its life-sustaining ability to bind, transport, and release oxygen in the body.

Hemocyanin transports oxygen in crustaceans, giving them a very different blood chemistry than mammals.

- But nature has gone far beyond just using iron ions to get the job done. For example, hemocyanin—rather than hemoglobin—transports oxygen in crustaceans like crabs and lobsters. This enzyme contains two copper ions coordinated to the protein matrix, giving lobsters, crabs, and many other marine animals a very different blood chemistry than mammals.

- Another of nature's chemists is the nitrogen-fixing Rhizobia bacteria that live in the root nodules of legumes like soy. The molecule that they produce that allows them to fix nitrogen is called nitrogenase, and it is a protein that contains not just one metal ion, but multiple clusters of them in complex with sulfur atoms.

- Scientists have discovered and characterized two different forms of nitrogenase, both of which rely on iron to function—one exclusively, and another in concert with molybdenum. Even though these metalloprotein structures have been well known for decades, their ability to accomplish the same reaction at room temperature and pressure that Fritz Haber could only accomplish at explosively high pressures and temperatures remains unexplained today.

- When someone finally unlocks the chemistry of these large, powerful organometallic molecules, the impact it will have on world food supply management will likely be as great as cold fusion would be for the power industry.

Suggested Reading

Lewis, D. E., "The Beginnings of Syntetic Organic Chemistry: Zinc Alkyls and the Kazan' School," *Bulletin for the History of Chemistry* 37, no. 1 (2002): 37–42.

McMurry, *Fundamentals of Organic Chemistry*, 7.3.

Wade, *Organic Chemistry*, 10.8, 10.9.

Questions to Consider

1. Propose two ways that you might use a Grignard reagent to prepare 3-methyl-3-hexanol from an alkyl halide and a ketone.

2. What is it about the atomic structure of gadolinium that makes it such a powerful magnet?

Metals in Organic Chemistry
Lecture 25—Transcript

Today I'm going to show you how to do the impossible. OK, that's a little bit dramatic, but not entirely inaccurate. Now, what would you say if I told you that I know how to deprotonate an alkane? Now remember, the hydrogen of an aliphatic alkane, like the hydrogens in a molecule of ethane, for example, are the single weakest acids of any compound that we've studied to date, with a pK_a of somewhere around 60. Now, that's literally trillions of times less acidic than alkenes, which are, themselves, trillions of times less acidic than alkynes, which are an extraordinarily weak acid in their own right.

But if we could deprotonate an alkane, just imagine what a powerful nucleophile the resulting carbon anion would be. It's a tempting target, isn't it? But how do we do it? So if we were to have a go at creating such a motif using proton transfer, we would need a base whose conjugate acid has a pK_a of at least 60, but, what would such a base be? Right, the only option would be deprotonated alkane, so we're right back at our original problem.

OK, you've got me here; this is a little bit of a trick question. It really is just about impossible to deprotonate an alkane under any kind of sensible laboratory conditions, but there is a class of compounds which can be made to act like carbanions. In order to do this, we have to abandon the safety and simplicity of the first few rows of the periodic table. We have to use a remarkable marriage between the usual suspects of organic chemistry, like carbon, nitrogen, and oxygen, and a newcomer to our discussion, metals.

Today we're going to investigate a small slice of a field known as organometallic chemistry. Now, before I continue, I need to advise you that we're only going to look at a very small set of organometallic compounds and reactions. Organometallic chemistry is used in a wide variety of applications, as we'll soon see, and it's a science worthy of its own graduate level course when considered in its entirety. Organometallic chemistry as a proper science got its start in a familiar place, the lab of Alexander Butlerov at the University of Kazan in 1800s Russia. Butlerov had noticed that dialkyl zinc reagents, or chemicals containing two organic alkyl groups bonded directly to a zinc atom through their carbons, could be used to produce alcohols from ketones by transferring their alkyl group to the carbonyl carbon of the ketone.

No doubt excited about the prospect of developing a new method for the formation of carbon-carbon bonds, Butlerov passed this project onto his protege, Alexander Zaitsev. Zaitsev was able to apply its chemistry to several new substrates, but always with limited success. These zinc-based reagents proved to be unpredictable and difficult to work with, and as the sun set on Zaitsev's career, he seemed no closer to developing reliable methods or rules for their use in synthesis. It would appear that this particular line of research was destined for obscurity.

Then, in 1900, a young French graduate student by the name of Victor Grignard blew the lid off of organometallic chemistry when he devised a powerful, versatile, and adaptable organometallic reagent for use in the reliable production of carbon-carbon bonds, which had eluded Butlerov and Zaitsev. Grignard's story is truly inspirational for anyone struggling with their schooling. He took a long and winding road through his college education, and after starting as a secondary education student, he failed to pass a standardized math examination, which landed him in the military for a short time. When he returned from his service, he managed to move ahead with his education, ultimately finding himself working in the lab of Philippe Barbier at University of Leon, in France, in the late 1890s.

It was there, in Barbier's lab at Leon, conducting experiments for his doctoral dissertation, that Grignard acted on a hunch, replacing the zinc from the only widely known class of organometallics of the day with what he believed would be more reactive magnesium. One of his first reactions was the combination of isoamyl bromide with magnesium, followed by carbon dioxide. And the reaction produced what Grignard himself described as, quoting here, "A colorless liquid distinctly of acidic character with an odor of perspiration and at the same time of butter." So it sounds like Grignard was really suffering for his science, having chosen a reaction which produces a fragrance reminiscent of sweaty butter. Nonetheless, I'm sure he was glad that he did, since his work not only earned him his doctorate, but one half of the Nobel Prize 12 years later. His co-recipient was his countryman Sabatier for his metal catalyzed hydrogen addition reactions.

So let's take a look at a Grignard reaction, in fact, one of the original Grignard reactions conducted by Victor Grignard himself. Now, this is a bit of an oversimplification, but it all holds true what's going on. Victor Grignard started with the reagent isoamyl bromide, which is alkyl halide. To this he added magnesium. Now, magnesium is two electrons shy of an octet, and so it would really like to have two bonds. So, when mixed with magnesium, isoamyl bromide creates this new compound where the magnesium has been inserted between the carbon and bromine atoms.

So the consequence of this is we've placed our metal next to a carbon. Now here's where organometallic chemistry really starts to differ from what we're used to seeing. Magnesium has a very low electronegativity, and because of that, it creates a very strong dipole with the negative end pointed towards the carbon. So this is very high electron density near the carbon to which the magnesium is bonded makes it act more like a carbanion, so we can think of it as being this, even though we know that this species is very difficult to generate.

So, Grignard mixed this Grignard reagent, which acts like a carbanion, with carbon dioxide, which has a very electrophilic carbon, the only carbon in the molecule, of course. So when we combine a very powerful nucleophile, which acts like a carbananion with a very good electrophile, such as carbon dioxide, we should all be well aware of what's going to happen next, nucleophilic attack creates a new carbon-carbon bond, and acidifying gives us our final product, isocaproic acid or isoamyl acid.

Of course, when Grignard made it, his reagent was not a carbanion, but an alkyl magnesium bromide reagent, which means that we also have to account for the magnesium in the bromine. Now, what happens to them during the nucleophilic attack is a bit complicated and case specific, but at the end of the reaction, the by-product is magnesium hydroxy bromide. So here we can see one of the driving forces behind the Grignard reaction. Notice that what started as a neutral magnesium atom is now a magnesium 2+ ion with a full octet. So powerful is the desire for magnesium to have that octet that it can be used to drive carbon-carbon bond formation.

Grignard continued his research and discovered that his new nucleophile could be used reliably to produce acids from carbon dioxide, secondary alcohols from aldehydes, and also tertiary alcohols from ketones. In fact, Grignards are extremely versatile, offering not only ways to make alcohols and acids, but also, aldehydes from amides, tertiary alcohol from esters, and many, many other reactions. It seems that when Grignard reagents are involved, it's pretty much open season on anything with a carbonyl group.

But there's much more to the Grignard reaction. Easy though it is to run, there are certain special considerations which must be taken to ensure its success, two, in particular. First, the Grignard reagent itself is only stable in ether solvents, like diethyl ether or tetrahydrofurn. Second, the reaction mixture has to be kept extraordinarily free of protic solvents, including water. So let's consider each of these two restraints individually, then collectively.

One of the great limitations of Grignard reactions is that alkyl magnesium halide reagents can only form in ether salts. But why is this? Well the answer lies in the ability of solvents to coordinate metals. Now, what I mean by this is that the solvent must have oxygen atoms with lone pairs which can stabilize the Grignard by donating electrons to form a stabilizing metal-oxygen interaction in solution. That alkyl magnesium halide motif is so polar that it needs a solvent which will be able to arrange its dipole beneficially to interact with the positive-charged density at the magnesium. For this purpose, ether solvents are just the ticket. Nonpolar solvents like hexanes or benzene are not capable of forming these kinds of stabilizing interactions, so they are clearly out as a solvent.

So Grignard reagents are polar; they're really polar; we get that. So why can't we use more polar solvents like ethanol to dissolve them for reactions? Certainly, ethanol has enough polar oomph get this reagent to form in solution, right? Well, it does, but with one devastating caveat. Unlike ether, ethanol is protic and has a pK_a of about 17. Remember, Grignards react like carbanions, whose conjugate acid has a pK_a about 60. So a Grignard reagent might form in ethanol for just a split second before reacting with that ethanol to form ethoxide and a hydrocarbon. Clearly, this is not a good way to make a Grignard. Now, ether like diethyl ether and tetrahydrofuran are special, because they strike a balance between their ability to solvate a Grignard with their inability to react with it, and this is why they have been the preferred solvent for this sort of reaction from 1900 right up until today.

But using ether and THF, tetrahydrofuran, isn't enough, because these solvents are slightly polar, so ether can dissolve a small bit of water itself. In fact, when left open, a bottle of ether will absorb about 1.5% of its own mass in water from the atmosphere. So a bottle of ether left open by careless colleague, or even simply one which has been opened and closed enough times to allow a small amount of air into it, can contain enough water to effectively ruin a Grignard synthesis. Our solution to this is to use ether which has been chemically dried for use, and we accomplish this through the use of inorganic drying agents, like magnesium sulfate, or special kinds of material called molecular sieves, which act like little mineral sponges, absorbing the water from the ether and trapping it as a solid complex, which we can simply filter or decant to separate.

Grignard reagents are incredibly an versatile tool for attaching new carbons to electrophiles, so, let's take a moment and summarize just a few of the tools in the Grignard toolbox and their limitations. First, Grignards are made using alkyl halide and magnesium metal mixed in an ethereal solvent, but they're not limited to aliphatic R groups. Commonly used Grignards include unsaturated groups also, like allyl magnesium bromide and phenyl magnesium bromide.

Grignard reagents provide a way to access the kind of reactivity we would expect to get from an aliphatic carbanion without the fuss of having to deprotonate one of the least acidic hydrogens known to man, the alkane C-H bond. Grignards can be reacted with carbon dioxide to form organic acids, with aldehydes to form secondary alcohols, and with ketones to form tertiary alcohols. They can attack secondary imines to form secondary amides with varying degrees of branching. Carbonyl compounds with leaving groups, like acid chloride attached to the carbon, can undergo multiple attacks by Grignard reagents, resulting in the addition of two alkyl groups, giving us an alternate route to make secondary or tertiary alcohols.

So far today, we've discussed the use of metals to activate hydrocarbons and turn them into nucleophiles. Of course, organometallic chemistry is far more varied and complex a subject than just this small window on the science would appear to imply. Organometallics have found their first few applications in the world of organic synthesis, but their real potential cannot be appreciated until we take a look at another one of their uses, catalysis.

The use of organometallics hearkens back to our old friend Sabatier, who developed a technique for hydrogenating fats using platinum metal, and who, incidentally, shared the Nobel Prize with Victor Grignard. In modern laboratories, the reduction of alkenes to alkanes is still a widely used technique, but the heterogeneous nature of Sabatier's method leaves a lot to be desired. Hydrogenating double bonds at the surface of a solid catalyst may work well on industrial scales, but in the laboratory, a reaction which takes place in solution is far more desirable.

If we had access to a homogeneous catalyst, one which could be dissolved in organic solvents along with substrates, we could run reactions much more efficiently and cleanly in the lab, wasting less reagents and amplifying our yields. But the problem with this is that metals themselves are not generally soluble in the kinds of solvents that one would use to dissolve an alkene. But in 1966, a British chemist by the name of Geoffrey Wilkinson published a paper outlining a way to solubalize a large transition metal, like rhodium, by bonding it to organic molecules which we call ligands. These ligands are specially designed to be strong Lewis acids, which donate stabilizing electron pairs to the metal ion, forming a complex which gives it the solubility characteristics of the ligands while retaining the catalytic activities of the metal.

Wilkinson's work is generally regarded as the seminal work in the field organometallic chemistry. His discoveries in this field launched an age of investigation into metal ligand interactions that continues to this day in labs around the world. Wilkinson's work has inspired the creation of some really exciting and exotic-looking complexes. You'll see extensions of his work in such life-changing compounds as contrast reagents used for MRI imaging. The magnetic properties of many large metals make them useful as tools to enhance the sensitivity of MRI imaging machines, but the problem is that most heavy metals are highly toxic by themselves or don't accumulate in the necessary tissues for them to be useful.

The gadolinium ion is one perfect example of this. Now when gadolinium is in its +3 state, it's a powerful atomic-sized magnet, and its presence is easily detected by such complex magmatic instruments as the MRI you may see in a hospital. Unfortunately, the gadolinium itself is very toxic and doesn't accumulate well where we need it. So, one solution to this problem is to complex metal ions with organic ligands, which render them far less toxic and also can cause them to selectively accumulate in areas of interest in the body, assisting in the imaging process. A classic example of this is the complex of gadolinium (III), which is marketed under the trade name Dotarem.

Gadolinium itself is an F-block metal, and its 3+ ion contains seven electrons filling its F subshell, its outermost subshell. Electrons like their space, so those seven outermost electrons in the F orbitals of gadolinium are unpaired. This gives it a tremendously powerful magnetization, so this tiny little ion of gadolinium is producing an extremely powerful magnetic field but only at atomic distances.

Dotarem is an ingenious way to exploit the magnetic properties of gadolinium ions likes this without dealing with toxicity issues and solubility problems. This is accomplished using a complex organic ligand, which looks like this. We call molecules with large rings of many atoms like this one macrocycles. The key to Dotarem's safety and efficacy is not only this macrocyclic ligand which sits on top of the metal, but it allows it to attach its carobxylates around the edges. These nitrogens and carboxylates contain lone-pair electrons in regions of dense negative charges, which coordinate to the gadolinium (III) ion. So, this forms a cage around the ion, preventing it from reacting with anything as it circulates in your bloodstream. The ligands also have the effect of changing the solubility characteristics of the complex, presenting an organic face instead of an ionic face. This helps the Dotarem accumulate in tumor cells.

The final piece of the puzzle is this last little bit of space on the bottom, which is complexed to the oxygen of a water molecule. We place our water molecule under there, you can see how close it is to the protective gadolinium ion. The proximity of the water molecule to that gadolinium (III) makes it very easy to detect using an MRI machine. Physicians can then use

MRI techniques to look for the magnetic effects of the gadolinium (III) on the water molecule inside the patient, thereby helping them locate and image problem areas.

Not to be outdone by modern chemists, nature herself has also devised ways to use organic molecules which help not just humans, but all other known species to get the benefits of the chemistry of metals while coping with their inherent toxicities. So, let's take a look at a subunit of the structure of human hemoglobin. It's pretty big molecule, isn't it? Buy if I remove all the protein atoms, we see that left over is a large, cyclic ligand, known as a porphyrin, and this porphyrin coordinates an iron atom in the center from four planar positions.

This metal ligand complex is commonly called heme, and is the namesake of the oxygen-carrying protein hemoglobin. Now, free iron is toxic in the blood, but it's the coordination of these four nitrogen atoms and also two histidines, which come in from the protein itself, which give this molecule its life-sustaining ability to bind, transport, and release oxygen in the body without being toxic to us.

Now, the heme ligand not only plays a role in the oxygen-carrying ability of hemoglobin, but its highly conjugated structure also makes it highly colored. Now, with the iron in place, the heme takes on a familiar deep red color that we associate with oxygenated blood, But when physical trauma causes cells in the body to rupture and die, the exposed heme from their hemoglobin has to be cleaned up, and this is done by enzymes, which first break open the ligand, and this breaking of the ring forms a new compound known as biliverden. It's a dark green colored compound. The biliverden is then further reduced into a compound known as bilirubin, which is yellow in color, and this is the process of cleaning up a porphyrin ligand, and it's also what we commonly associate with the process of bruising. So, the colors that you see underneath the skin are actually the free heme from the traumatized cells being cleaned up by your own body.

But nature's gone far beyond just using iron ions and porphyrin ligands to get jobs done. Another great example is hemocyanin. Rather than hemoglobin, hemocyanin still transports oxygen; it does this in crustaceans, crabs, and lobsters, those sorts of animals, but it does so using two copper ions coordinated directly to a protein matrix, which gives these lobsters, crabs, and other marine animals a very different blood chemistry than mammals. So in this case, the protein itself is acting as the ligand for the copper ions which carry oxygen.

Let's look at one more example, vitamin B12. Now, vitamin B12 includes a compound called cobalamin, and as the name implies, cobalamin has a cobalt ion complexed in its center. Now, this vitamin is used in the body to conduct essential functions, including synthesis of folic acid, and again, it's all dependent on having the ligand in place to coordinate that metal.

So nature one-ups the organic chemists once again, having devised ligands far more complex than those of Wilkinson's catalyst or even Dotarem, and they're capable of conducting much more powerful directed chemistry. But why stop at one metal? Yet another of nature's chemists who continues to embarrass us all is the nitrogen-fixing rhizobia bacteria, which live in the root nodules of legumes, like soy. The molecule that they produce, which allows them to fix nitrogen, is a protein called nitrogenase, and it contains not just one metal ion, but multiple clusters of them, in complex with sulphur atoms.

Scientists have discovered and characterized two different forms of nitrogenase, both of which rely on iron to function, one of them exclusively, and another one in concert with molybdenum. Now, even though these metaoprotein structures have been well known for decades, their ability to accomplish the same reaction at room temperature, and pressures, that Haber could only accomplish at extremely high pressures and temperatures, remains unexplained even today.

Now, when someone finally unlocks the secret to the chemistry of these large, powerful, organometallic molecules, the impact it will have on the world food supply is likely to be as great as cold fusion would be for the power industry.

So today we started with one of Butlerov's projects in the 1800s. We took a look at how he and Zaitsev conducted some of the very first organometallic experiments. Then we spent some time covering the reaction developed by Victor Grignard in 1900, which bears his name, and won him the Nobel Prize, and also breathed new life into floundering science of organometallic chemistry. We talked about how his alkyl magnesium halides have the properties of carbanions, including strong nucleophilic character and extreme basicity. We covered how this nucleophilic character makes no carbonyl safe in the presence of a Grignard, leading to a large library of potential substrates, which can be used in conjunction with them.

We saw how Geoffrey Wilkinson is usually regarded as the father of modern organometallic chemistry, having devised a method to get rhodium and platinum of Sabatier's heterogeneous catalysts into solution, making them more effective homogeneous catalysts. We then advanced on Wilkinson's ideas and took them into the biological realm, seeing how compounds, like Dotarem, get highly magnetic but extremely toxic gadolinium ions safely into the bloodstream of cancer patients, where they can help to detect tumors.

Finally, we indulged mother nature a bit more and took a look at some pretty representations of metal-containing proteins like hemoglobin, and hemocyanins, nitrogenases, and some vitamins, all of which carry out spectacularly focused chemistry in one way or another helping to sustain the lives of those creatures that use them.

We've come a long way from the simplest of all organic compounds, methane, to organometallics the size of nitrogenase. We've nearly run the gamut of molecular size, but there is one more class of organic compound to consider, the undisputed king of molecular size, polymers. And they're the last class of organic compounds on our list.

Next time we're going to take a look at some world-changing polymers, how they were discovered, how they're made, and how their size and modular composition make them an extremely versatile class of compounds. I'll see you then.

Synthetic Polymers
Lecture 26

A tremendous amount of attention has been poured into understanding, designing, and creating useful polymers. From the tires on your car, to the man-made fibers in some clothing, to medical equipment, body armor, and more, the importance of synthetic polymers to our modern lifestyles cannot be overstated. This lecture will provide a short survey of polymer chemistry—how polymers are designed, how they are made, and a few ways in which they have changed the world.

Polymers
- Polymers are defined as large molecules consisting of one or more repeating units known as monomers. Biologically relevant polymers include DNA, RNA, proteins, and starches.

- Synthetic polymers are classified in more than one way, but a typical method is to distinguish them based on how they are prepared. From this perspective, most polymers fall into one of two general classes: addition polymers or condensation polymers.

- Addition polymers are formed when monomers react with the end of a growing chain. Addition polymers tend to be prepared from alkene subunits, trading less-stable pi bonds for the new sigma bonds that link them together. The first wholly synthetic polymer, polystyrene, falls into this class of compounds. The process of creating addition polymers consists of three phases: initiation, propagation, and termination.

- In the case of polystyrene, polymerization reaction initiates when a sample of styrene is heated in the presence of a small amount of benzoyl peroxide. The heat causes the benzoyl peroxide to dissociate into two phenyl radicals, a process encouraged by the loss of carbon dioxide gas.

- In the second step—propagation—as each monomer encounters the reactive radical at the end of a chain, a new sigma bond is formed homogenically, regenerating a radical at the end of the lengthened chain.

- In the case of polystyrene, this process repeats itself several thousand times before the third phase, termination, takes place. In termination, two growing radicals react with one another, forming one last sigma bond that consumes the radicals without generating a new one.

- Addition polymers are very attractive targets for commercial materials because their properties can be tuned by controlling the overall size of the macromolecules produced in the chain-growth reaction. The size of addition polymer molecules can be controlled simply by altering the relative amount of benzoyl peroxide added: Relatively fewer benzoyl peroxide molecules means fewer growing chains, and because growth can only take place at the end of an initiated chain, one expects to obtain longer polymers in such a scenario. Additionally, fewer growing chains equates to a lesser chance that two growing chains will encounter one another in a termination step.

- Another example of an addition polymer is the polyethylene used in plastic bottles, formed by the addition of ethene monomers to one another. Polyvinyl chloride (PVC), most well known as a substitute for metal plumbing pipes, is assembled in a similar reaction using vinyl chloride monomers. Even the once-popular cookware coating Teflon is formed by polymerization of tetrafluoroethylene monomers.

- The second major class of polymers is condensation polymers, which differ from addition polymers in that their growth is driven not by trading of pi bonds for sigma bonds, but by the formation of small molecule by-products.

- A perfect example of this is the formation of nylon-6, in which molecules of 6-aminohexanoic acid are joined together in a reaction that produces not only an amide union between monomers, but also a molecule of water as a by-product.

- Condensation polymers are also called step-growth polymers, because unlike chain-growth polymers, condensation can take place at any time between any monomers or growing polymers.

Copolymers

- When polymers are constructed from multiple monomers, we refer to them collectively as copolymers. Copolymers can take on a vast array of arrangements, including alternating copolymers, in which monomers regularly repeat; random copolymers, in which a statistically random arrangement of monomers exists; block copolymers, which are characterized by large defined regions consisting of one monomer type or the other; and graft copolymers, in which a polymer of one monomer supports polymers of another.

- A familiar alternating polymer is produced when vinyl chloride and vinylidine chloride react in a radical chain-growth polymerization. In this reaction, monomers add in an alternating order, with vinyl chloride monomers reacting at initiated viniylidine termini and vinylidene chloride monomers reacting at initiated vinyl chloride termini. The result is an alternating polymer called saran, which is used to make the very popular food wrap that bears its name.

- But alternating polymers can just as easily be produced using condensation polymerizations. One example of this is the polymer nylon-6,6, which is prepared using equal parts of adipic acid and 1,6-hexanediamine.

- By producing nylon in this way, the orientation of alternating amide bonds is reversed from those in nylon-6. In fact, it is more accurate to say that the amide bonds in nylon-6 are reversed, because nylon-6,6 was actually the first of these to be patented by Dupont in 1935.

- The development of the homopolymer nylon-6 by BASF was an attempt to produce a material with similar properties that did not infringe on Dupont's intellectual property. Both of these polymers were a huge success and are still used today as the backbone for some of the most modern materials produced by both companies.

- Random copolymers are produced by mixing monomers, but unlike alternating copolymers, growing random copolymers can incorporate any of the monomers in any order, leading to a statistically random distribution of monomers in the macromolecule.

- An example of a random copolymer is polybutyrate, which is a polyester compound built from two distinct subunits. The distinction is that these subunits are randomly oriented throughout each polymer strand. Incredibly, polybutyrate is fully biodegradable. Plastic bags and wraps produced from this flexible random copolymer are completely deconstructed by moisture and microbes in the ground.

- One of the more exciting advances in recent decades has been the development of block copolymers, which aim to exploit the best properties of more than one polymer by linking together large regions of each, called domains.

- The block copolymer SBS—which stands for poly(styrene-butadiene-styrene)—is used in many modern applications, including tires and shoe soles. In a single molecule of this block copolymer, a domain of polybutadiene is sandwiched in between two polystyrene domains.

- The polystyrene blocks give this polymer great toughness, while the polybutadiene domain becomes softer under high-temperature conditions, allowing the material to be more easily moldable like softer rubbers. The result is a material that behaves more like plastic at high temperature, but more like vulcanized rubber when cooler.

- Graft copolymers are characterized by a uniform backbone polymer that is decorated by strands of a different polymer attached covalently along its length. In many cases, graft copolymers can be designed to have certain properties of both polymers that make it up.

- Examples include long polysacharides like cellulose onto which other polymers are grafted. This has the effect of producing a very large molecule with many of the properties of the smaller grafted polymers while retaining the biodegradability of the cellulose backbone.

Cross-Linking Polymers

- Natural rubber is a somewhat complex mixture created by trees to defend themselves from wood-eating insects. It has quite a few minor components, but the big player in its chemical properties is a polyisoprene macromolecule made on long, unsaturated hydrocarbons, making it very thick and insoluble in water.

- The problem is that these polyisoprene macromolecules are only held together by dispersion forces stemming from their electron clouds. These forces are enough to make the molecule viscous, but they can be overcome. So, when we pull on natural rubber molecules, they can slide back and forth with one another.

- In the latter part of the 1830s, Charles Goodyear invented a method for producing a variation of rubber that would be strong enough to produce objects that could be molded into a desired shape but would then retain that shape across a greater range of conditions. His method involved steam heating a mixture of natural rubber with sulfur.

- What Goodyear accomplished by the addition of sulfur was to cross-link the polymer molecules with covalent bonds to sulfur. With these bonds in place, the material can still be deformed slightly, but as soon as it is released, it snaps back into its original position.

- So, the presence of these cross-links is what gives vulcanized rubber its toughness and ability to retain its shape—exactly what you would hope for in a product like an automobile tire. It also, unfortunately, makes the material very difficult to deconstruct, meaning that vulcanized rubbers are difficult to recycle.

© Fuse/Thinkstock.

Modern tires can be deformed slightly, but will snap back into their original position.

Polycarbonates and Bisphenol A

- Another common polymer motif is polycarbonates. These polymers are found in materials as tough and rigid as bulletproof glass and as soft and supple as commercial water bottles. Both of these products are made from the condensation of carbonic acid with bisphenol A (BPA) to form an alternating copolymer.

- Recently, concern has arisen over the possibility that these polymers can hydrolyze back into their starting materials over time, leaching bisphenol A into the water or beverage that they contain. The reason for so much concern is mounting evidence that bisphenol A imitates estradiol in the body, potentially causing developmental and reproductive problems.

- Some studies suggest that BPA is less than 0.0001 as powerful of a hormone as estradiol is, suggesting that this controversy may be overblown. Still, we simply don't know what the long-term effects of exposure to low levels of BPA might do. Regardless, many people are sufficiently wary of BPA that they are willing to purchase products labeled "BPA-free" without giving much thought to what is actually in the plastic.

- The polymers industry has responded to these concerns by offering BPA-free plastics with a great deal of fanfare. What they do not tell you, however, is that most of these products are still polycarbonates that are simply made using molecules like bisphenol S or bisphenol F in place of bisphenol A in an analogous reaction.

Most products labeled "BPA-free" are still polycarbonates.

Plasticizers

- Another hot-button plastics issue is the use of compounds called phthalates. Phthalic acid is a small organic diacid that is easily esterified with a number of aliphatic alcohols to produce small molecules called phthalates.

- Phtalates—like dibutyl phthalate, for example—are sometimes added to polyvinyl chloride to soften the plastic enough to make it useful over a wider range of applications. Plasticizers work by disrupting the packing of the polymer chains—in this case, in the PVC, causing it to soften. That new car smell many people love is caused by these plasticizers, escaping the newly prepared plastics into the air.

- Polyethylenetetraphthalate (PETE) is used in food packaging and does not contain isolated phthalate molecules. The "-phthalate" portion of PETE's name refers to the monomer that is condensed to prepare the polymer. So, in this case, the phthalate is locked securely in the matrix of this nonbiodegradable plastic, where it cannot easily be released.

- So, PETE doesn't release phthalates into the environment even though "-phthalate" is in its name, and PVC very often does release phthalates, even though the term "-phthalate" is nowhere to be found in its name.

Suggested Reading

McMurry, *Fundamentals of Organic Chemistry*, 15.3, 15.4.

Wade, *Organic Chemistry*, Chapter 26.

Questions to Consider

1. How will increasing or decreasing the amount of initiating reagent used in chain-growth polymers affect the chain size in the completed polymer molecules?

2. Shown below is the structure of the well-known copolymer Kevlar. Propose two monomers that might be used to produce Kevlar using a condensation reaction.

Synthetic Polymers
Lecture 26—Transcript

In the latter part of the 1830s, a desperate businessman by the name of Charles Goodyear was near the end of his rope. Having already failed in the hardware business in Philadelphia, he had turned his attention to a newly available material from Brazil, called rubber. Like many other entrepreneurs at his time, Goodyear believed that this pliable, moldable substance had enormous potential to be put to work in a number of potentially profitable applications, but there were some problems.

Despite its usefulness in waterproofing applications, rubber itself was very soft, pliable, and tended to deform plastically, meaning that once pulled or otherwise misshapen, it retained the new shape. Now, Goodyear had worked with natural rubber for about a half a decade, seeking out ways to strengthen it so that it can be used to produce objects which could be molded into a desired shape but would then retain that shape across a great range of conditions.

As seems to usually be the case with great discoveries, Goodyear's breakthrough was reached quite by accident. There are several variations of the story, but the most inspirational is this. Being nearly bankrupt, Goodyear had attempted to remove paint from one of his rubber samples for reuse. To do this he tried using nitric acid. He noticed that after removing the paint with nitric acid, the properties of his rubber sample had changed. He realized then that treatment with various other materials might alter the structure of rubber, or even become incorporated into it, changing its properties for the better.

Now, Goodyear wasn't much of a chemist, so he had to rely on blind trial and error to concoct his modified rubber materials. Being readily available and inexpensive, one of the modifiers he tried was elemental sulfur. Now, as the story goes, Goodyear walked into the general store in the town of Woburn, Massachusetts, where he was then living. It was then that he accidentally dropped a sample of his latest sulfur-treated rubber concoction onto a pot-bellied stove. In the ensuing charred mess, he noticed that some of that material's properties had changed.

Goodyear later worked out a more controlled process for producing the material he had first observed in the aftermath of his accident, and he finally settled on a method for production by steam heating a mixture of natural rubber with sulfur. His new material was first used by his brother-in-law to induce the then-sought-after ruffles in the fabrics of men's shirts.

Now, it turns out that Goodyear wasn't only limited in his chemistry skills; he didn't fare too well in the world of business either. Goodyear managed to obtain an American patent for his new material, but he failed to seize the same opportunity in the British market. It was an English entrepreneur and a competitor of Goodyear's by the name of Thomas Hancock who reverse engineered his invention from one of Goodyear's own samples and obtained the British patent. Shortly thereafter, one of Hancock's associates coined the term vulcanized rubber, naming the material for the Roman god of fire. Perhaps some would say that Hermes, Roman god of thieves, would have been a better namesake.

So, vulcanized rubber became not just a world changing material from which many durable rubber products could be made, but also, one of the first artificially modified polymers in human history. Polymers are defined as a very large set of molecules consisting of repeating units known as monomers. We've already looked at some biologically relevant polymers including DNA, RNA, proteins, and starches. But Goodyear's invention is noteworthy because it may have represented the first semi-synthetic polymer created by man. But it wasn't long after that that a German apothecary by the name of Eduard Simon managed to create a substance with similar properties not by modifying an existing natural polymer, but from individual molecules of styrene heat-treated to induce polymerization.

Now, by no fault of his own, Simon lacked the insight to fully understand what he had just done. He only new that heating styrene led to a material with properties very similar to rubber. This observation went largely unnoticed. To those who did notice, the simplest explanation was that he had created a conglomerate of small molecules held together in aggregates by intermolecular forces. After all, the creation of a covalent molecule with thousands upon thousands of atoms was not a reasonable explanation. That could never happen. Right?

It was nearly a century later that Simon's countryman, Hermann Staudinger, finally elucidated the structure of Simon's creation. It took some outside-of-the-box thinking to finally understand Simon's material. Staudinger finally postulated publicly that the heated styrene molecules had reacted with themselves to create immense molecules of great size. Staudinger's initial efforts to reach this idea, or to research this idea, were met with a remarkable skepticism. He found it difficult to secure the support of his colleagues, who all believed rubbers were aggregates of small molecules stuck together by some messy non-covalent force. Even the eminent Hermann Emil Fischer went on record against Staudinger's theories. But Staudinger stood fast, and it's a good thing that the did. His theory proved true. And having fought the good fight to prove it, he received his Nobel Prize in 1953. Ultimately, it was realized that styrene molecules had been joined together through their vinyl groups, creating carbon chains with hundreds or even thousands of atoms. But the benefit of 100 years of additional insight and his willingness to continue supporting his theory in the face of substantial opposition, Staudinger's theory of polymers, which he referred to as macromolecules, ultimately had proven true.

Now, since the days of Staudinger's work a tremendous amount of attention has been poured into understanding, designing, and creating useful polymers. From the tires on your car to the man-made fibers in some clothing, to medical equipment, body armor, food packaging, and even more, the importance of synthetic polymers to our modern lifestyles simply cannot be overstated. So, in this lecture, we're going to take a short survey of polymer chemistry, how polymers are designed, how they're made, and a few ways in which they've changed the world.

Synthetic polymers are classified in more than one way, but a typical method is to distinguish them based upon how they're prepared. From this perspective, most polymers fall into one of two general classes, addition polymers, also called chain-growth polymers, or condensation polymers, sometimes called step-growth polymers. So, let's take a look at each of those classes one the time.

The first class of polymers based upon their synthesis are addition polymers, also called chain growth. Now, any polymer will be constructed from a set of small molecules known as monomers. Now, I've represented my monomers here as simply these red objects that are going to do the chemistry that we would expect of a molecule, which is very small, but is a monomer.

To this we're going to add another type of molecule called an initiator. That's because in the case of addition polymers, my individual monomers are not reactive in and of themselves. But, in the presence of an initiator, my monomer can react. And only when attached to the chain will that new monomer become reactive again, sequestering more and more the monomer to the chain. In short, addition polymers grow from one end to the other in a fashion much like this. So, their sizes are very predictable and very easy to manipulate. A classic example of an addition polymer are those formed from alkene subunits, trading less stable pi bonds or the new sigma bonds which link them together. The first entirely synthetic polymer, polystyrene, falls into this class of compounds.

The process of creating addition polymers consists of three phases—the initiation, the propagation, and the termination. In the case of polystyrene, polymerization reaction can be initiated when a sample of styrene is heated in the presence of just a small amount of benzoyl peroxide. Now, the heat causes the benzoyl peroxide to dissociate into two fennel radicals, a process encouraged by the loss of carbon dioxide gas. In the second step, propagation, as each monomer encounters the reactive radical at the end of the chain, a new sigma bond is formed homogenically, meaning one electron from each side. This regenerates the radical at the end of the lengthening chain and perpetuates the reaction. In the case of polystyrene, this process repeats itself several thousand times before the third phase, termination, takes place. In termination, either two growing radicals react with one another, or something of that like, forming one last sigma bond, which will ultimately consume the radicals without generating a new one.

Addition polymers are very attractive targets for commercial materials, because their properties can be turned or tuned by controlling the overall size of the macromolecules produced in the chain-growth reaction. For example, the size of the polystyrene polymer molecules can be controlled simply by altering the relative amount of benzoyl peroxide initiator, which is added. Relatively fewer benzoyl peroxide molecules means fewer growing chains.

And because growth can only take place at the end of an initiated chain, one expects to obtain longer polymers in this kind of scenario. Additionally, fewer growing chains means less likelihood that two growing chains will encounter one another in a termination step.

Some other examples of addition polymers include polyethylene, used in plastic bottles, formed by the addition of ethene monomers to one another. Polyvinyl chloride, or PVC, most well known as a substitute for metal plumbing pipes is assembled in a similar reaction using vinyl chloride monomers. Even the once-popular cookware coating Teflon is formed by polymerization of tetrafluoroethylene.

Now let's take a look at the second class of polymers, condensation polymers, also sometimes called step-growth polymers. This has become a very popular class of polymers, because they're relatively easy to make. Condensation polymers form, again, through conjoining more than one, obviously, monomers. Right? We want long chains, so we need lots and lots of monomers. But in the case of condensation polymers, each monomer itself is reactive without any initiator necessary. So in the case of condensation polymers, rather than having an initiator dictate when and where my monomers can join together, my monomers can really join together anywhere at any time. So in this case, I have a growing set of ever larger and larger polymers forming as my monomers become dimers, which can condense with other monomers or dimers, and so on and so on.

But you'll notice that in this process I get a wide variety of lengths in my polymers as they grow. So although condensation polymers are very easy to make, they are very difficult to control from a polymer length standpoint. Now generally, condensation polymers differ from addition polymers in that their growth is not driven by trading of pi bonds for sigma bonds but by the formation of small molecule by-products, like water or hydrochloric acid. A perfect example of this is the formation of nylon 6 in which molecules of 6-aminohexanoic acid are joined together in a reaction which produces not only an amide union between monomers, but also a molecule of water as a by-product.

Now, so far in this lecture, I've shown examples of polymerization reactions in which only one type of monomer is used, leading to very simple, repeating structures of that monomer. Now, make no mistake, polymers like polystyrene, PVC, and nylon 6 are extremely valuable and useful materials. But we can significantly increase the complexity and, therefore, the tunability of polymers if we use more than one monomer in the process.

When we use more than one monomer to create a polymer, we call the class of compounds created copolymers. So, the examples I'm going to show here are using only two different monomers, but we could just as easily have three or four or any number of different monomers. Now, copolymers can be constructed using either one of our two major mechanisms for building a polymer, so they're not necessarily chain-growth or step-growth polymers. But what is true is that when I have a collection of monomers like this one, where I have indicated one monomer in red and one monomer in blue, they can combine in several different geometric relationships, each of which can lead to a polymer with different properties.

Possibly the simplest example of this is what we would call an alternating copolymer in which each monomer is spaced with two of the other monomers around it, so red, blue, red, blue, red, blue, all the way across. But a second arrangement would be what we would call a random copolymer in which the order of each monomer within the overall chain is completely statistically random. A third potential arrangement is what we call a block copolymer. In a block copolymer, long stretches of a certain type of monomer exist which are then punctuated by long stretches of the other monomer or monomers within the copolymer.

And as one last example of a type of structure that copolymers can form, we have something called a graphed copolymer, in which one of the monomers serves to form the backbone while the other decorates that backbone by branching away from it. So, an alternating polymer that we're familiar with is produced when vinyl chloride and vinylidene chloride react in a radical chain-growth polymerization. In this reaction, monomers add in an alternating order, with vinyl chloride monomers reacting at initiated vinylidene termini, and vinylidene chloride monomers reacting at initiated vinyl chloride termini. The result is an alternating polymer called saran.

This is the polymer used to make the very popular food wrap which bears its name. Isn't that remarkable? Just one additional chlorine atom on every other monomer unit and we go from PVC, which is used in plumbing, to Saran Wrap, which is very flexible and is used as a food wrap.

But alternating polymers can just as easily be produced using condensation polymerizations. One example of this is the polymer nylon 6-6, not to be confused with nylon 6. Nylon 6-6 is prepared it using equal parts of adipic acid and 1-6 hexanediamine. By producing nylon in this way, the orientation of alternating amide bonds is reversed from those in nylon 6. In fact, it would be more accurate to say that the amide bonds in nylon 6 are reversed, since nylon 6-6 was actually the first of these to be prepared and patented by DuPont in 1935. The development of the homopolymer nylon 6 by BASF was an attempt to produce a material with similar properties which didn't infringe on DuPont's intellectual property. In fact, both of these polymers were a huge success and are still used today as the backbone for some of the most modern materials produced by both companies.

Random copolymers are produced by mixing monomers. But unlike alternating copolymers, growing random copolymers can incorporate any of the monomers in any order, leading to a statistically random distribution of monomers in the molecule. An example of a random copolymer is polybutyrate, which is a polyester compound built from two distinct sub units. Now, the distinction here is that these sub units are randomly oriented throughout each polymer strand. But what's really special about polybutyrate isn't how it's made, but the fact that its ester bonds make it fully biodegradable. Plastic bags and wraps produced from this flexible, random copolymer are completely deconstructed by moisture and microbes in the ground, making it a popular choice for disposable items of this kind.

Now one of the more exciting advances in more recent decades has been the development of block copolymers, which aim to exploit the best properties of more than one polymer by linking together large regions of each, which we call domains. For example, the block copolymer SBS is used in many modern applications, including tires and running-shoe soles. SBS stands for polystyrene butydine styrene. So, in a single molecule of this block copolymer, a domain of polybutydeine is sandwiched in between two polystyrene domains.

The polystyrene blocks give this polymer great toughness, while the polybutydeine domain becomes softer under high temperature conditions, allowing the material to be more easily moldable, like softer rubbers. The result is a material which behaves more like a plastic at high temperature, but more like vulcanized rubber when cooler, a perfect recipe for manufacturing tires. I think Mr. Goodyear would be proud.

Our fourth class, graphed copolymers, are characterized by a uniform backbone polymer, which is decorated by strands of a different polymer attached covalently along its length. In many cases, graphed copolymers can be designed to have certain properties of both polymers which make it up, for example, long polysaccharides, like cellulose, onto which other polymers are grafted. This has the effect of producing a very large molecule with the properties of those smaller grafted polymers, but retaining the biodegradability of the cellulose backbone.

And while we're back on the subject of Mr. Goodyear, let's take a look back at his creation knowing what we do now about polymers. Natural rubber is a somewhat complex mixture created by trees to defend themselves from wood-eating insects. It has quite a few minor components, but the big player in its chemical properties is this—a polyisoprene macromolecule made on long, unsaturated hydrocarbons, making it very thick and insoluble in water. Now, the problem in Goodyear's days is that these polyisoprene macromolecules are only held together by intermolecular dispersion forces stemming from their electron clouds. So let's look at this. These forces are enough to make the molecule viscous, but they can be overcome. So when I pull on my natural rubber molecules, they can slide back and forth on one another.

Now, what Goodyear had accomplished by the addition of sulfur was to cross link the polymer molecules with covalent bonds to sulfur. So, let me pull that polymer back together here. So, some of these pi bonds within the rubber were replaced by a sulfur bridge. So we'll put it a few sulfur bridges. And now let's pull on that polymer again and see if it deforms. I can deform the polymer, but as soon as I release it, it snaps back into place. This is because of the bridging sulfurs.

So, the presence of these cross links is what gives vulcanized rubber its toughness and ability to retain its shape, exactly what you would hope for in a product like an automobile tire. It also, unfortunately, makes the material very difficult to deconstruct, meaning that vulcanized rubbers are very difficult to recycle. Now this is the reason that tire yards exist. For many decades, non-biodegradable, non-recyclable vulcanized-rubber tires were simply piled up in huge tire yards with no obvious fate. Today, we do our best to reuse vulcanized rubber by cutting or chipping it into smaller pieces for use in paving projects and other applications. But melting and reforming vulcanized rubber is simply not possible. So truly recycling a tire by making it into another tire is not a viable strategy.

Yet another common polymer motif is polycarbonates. These polymers are found in materials as tough and rigid as bulletproof glass and as soft and supple as commercial water bottles. Both of these products are made from the condensation of carbonic acid with bisphenol A to form an alternating copolymer. Recently, concern has arisen over the possibility that these polymers can hydrolyze back into their starting materials over time, leeching bisphenol A into the water or beverage that they contain. The reason for so much concern is mounting evidence that bisphenol A imitates estradiol in the body, potentially causing developmental or reproductive problems in those who are exposed to it.

Some studies suggest that BPA is less than one ten thousandth as powerful of a hormone as estradiol is, suggesting that this whole controversy may be overblown. Still, we simply don't know what the long-term effects of exposure to low levels of BPA might do. No one is sure exactly what the risks are, but one thing is undeniable, many people are sufficiently wary of BPA that they are willing to purchase products labeled BPA free without giving much thought to what's actually in the product.

The polymers industry has responded to these concerns by offering BPA-free plastics with a great deal of fanfare. Sounds great, doesn't it? But what they don't tell you is that most of these products are still poly carbonates, which are simply made using molecules like bisphenol S, or bisphenol F, in place of bisphenol A in an analogous reaction. Now, if there is concern over a molecule like bisphenol A imitating estradiol, something tells me it's only a matter of time until bisphenol S, F, and others come under similar scrutiny, putting us right back where we were before.

One last hot-button plastics issue that you may have heard of is the use of compounds called phthalates. Phthalate acid is a small, organic diacid which looks like this. It's easily esterified with a number of aliphatic alcohols to produce small molecules called phthalates. Now phthalates, like this dibutyl phthalate, are sometimes added to polyvinyl chloride to soften the plastic enough to make it useful over a wider range of applications. Now, plasticizers work by disrupting the packing of polymer chains, in this case in the PVC, thereby, softening it. That new car smell that so much of us enjoy is caused by these plasticizers escaping the newly-prepared plastics into the air.

Now, it's worth mentioning that polyethylene terephthalate, commonly called PETE, is used in food packaging and does not contain isolated phthalate molecules at all. The phthalate portion of PETE's name refers to the monomer which is condensed in the prepared polymer. So in this case, the phthalate is locked securely in the matrix of this non-biodegradable plastic, where it can't easily be released. So, PETE doesn't release phthalates into the environment, even though phthalate is in its name. And PVC very often does release phthalates, even though the term phthalate is nowhere to be found in its name.

So in this lecture, we grazed the surface of nearly 200 years of synthetic polymer chemistry. Starting with the accidental vulcanization of rubber by Charles Goodyear during a clumsy moment in 1830s Massachusetts. We talked about the two major classes of polymerization reactions, addition and condensation, also called chain-growth polymerization and step-growth polymerization, respectively. We saw how step growth usually involves the reaction of alkene sub units with an activated growing polymer chain until termination takes place. Specifically, we took a look at radical polymerization, which is initiated by the dissociation of a small bit of benzoyl peroxide to initiate chain growth. By contrast, condensation polymers tend to proceed with the formation of small molecules as monomers, dimers, trimers, tetramers, and more and condense with one another forming ever-larger chains until the reagents are exhausted.

We then took a different view on polymers, classifying them not based upon the reaction which makes them, but the morphology of the polymer chains themselves, including homopolymers and the many layers of complexity added when copolymers come into play. We saw the alternating copolymer saran and the random copolymer polybuterate, both of which blend monomers closely together, usually forming a polymer with totally new properties. Then we considered how graphed copolymers and block copolymers can preserve certain properties of each homopolymer from which they are derived, helping chemists to design polymers perfectly suited to an intended application. Finally, we looked at a few recent concerns regarding popular polymers, including the estrogen-emulating bisphenol compounds used in polycarbonates and the use of plasticizers, likes phthalates.

Of course, what we saw in this lecture is just a small slice of polymer chemistry, a field made incredibly diverse by not only the vast library of monomers and linking motifs, but by the sheer titanic size of the molecules involved.

So macromolecules bring us to a fitting end to our survey of organic compounds and their reactions. In our next few lessons, we're going to focus not on how new compounds are made and put to use, but on how they are identified. Organic chemists use a very special set of techniques to be their eyes as they navigate the long and sometimes complex road to an effective synthesis, and no organic chemistry course is complete without it.

We'll begin to develop an understanding of the techniques that have been used to verify so much of the information that I've already given you. From identifying specific functional groups to discerning regioisomers and enantiomers with small molecules to determining full protein structures, spectroscopy is the tool that lets us observe the fruits of our labor in the lab. It's an exciting science in its own right, and it will be the topic of our next few lectures. And I'll see you for those soon.

UV-Visible Spectroscopy
Lecture 27

This lecture will discuss the techniques that organic chemists and many others use to identify the products of their work. Specifically, this lecture will explore the science of spectroscopy. Most simply put, spectroscopy is the observation of the interaction of light with matter. Over the next few lectures, you will learn about some of the ways that scientists coax structural information from molecules using spectrometers to sense how they interact with different forms of light. This lecture begins in the visible and ultraviolet regions of the spectrum.

Light

- What we generally refer to as "light" is, in fact, electromagnetic radiation, a form of self-propagating energy composed of perpendicularly oscillating electric and magnetic fields. Exhibiting both wave- and particle-like properties, light carries energy and momentum. It is the wave properties of light and the energy it carries that are most interesting to spectroscopists, and we therefore tend to classify light based on these properties.

- One of the properties associated with waves is the wavelength—the distance from one point on the wave to the next analogous point. Another property of light is velocity. Light has a fixed speed at which it travels in a given medium, regardless of its wavelength. Light in a vacuum always travels at 3.0×10^8 meters per second. The frequency of light is the number of equivalent points along the wave that pass by a given location in a specified time.

- The constant speed of light becomes a proportionality constant for the relationship between wavelength and the reciprocal of the frequency: $c = \lambda v$. To put it simply, because of the constant speed of all light, cutting the wavelength in half results in a doubling of the frequency.

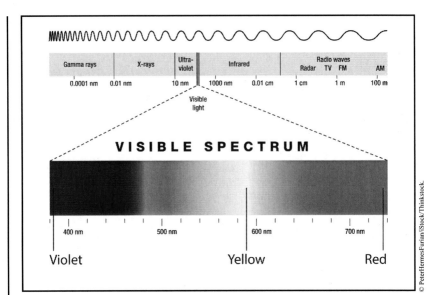

The electromagnetic spectrum includes not only what we can see with the naked eye, but also higher- and lower-frequency wavelengths.

- Another property of light that is crucial to understanding spectroscopy is that light does still have some particle character. We call these particles of light photons. A single photon of light contains an amount of energy that is directly related to its frequency by a factor called Planck's constant (h). A single photon of light has an associated wavelength, frequency, and energy that are all related to one another.

- We classify light according to its frequency or wavelength, which are related to the amount of energy carried by a single photon of that light. The known range of wavelengths is called the electromagnetic spectrum. This spectrum includes, in order of increasing frequency (and energy): radio, microwave, infrared, visible, ultraviolet, X-ray, and gamma ray.

The UV-Visible Spectrometer

- In 1801, Wilhelm Ritter acted on a hunch that there was more to the spectrum than what Sir Isaac Newton had seen with his own eyes more than 100 years earlier. The problem is that Wilhelm was a human, and his eyes couldn't see any farther along the spectrum than Newton's could. Ritter needed a different kind of eye, and he had just the tool for the job—silver chloride.

- Silver chloride is an inorganic compound that is most famous for its use in photographic film. Solutions of silver chloride undergo a chemical reaction in the presence of visible light that causes them to dissociate into a suspension of silver metal and chlorine gas. In other words, the solution turns dark on exposure to light.

- It was known even before Ritter that this reaction did not proceed at the same speed in different colors of light. He could disperse light with a prism, then place vials of silver chloride solution in different parts of the resulting spectrum, and clearly see that the reaction was fastest in blue, slower in green, and slower still in red.

- Ritter's real contribution was that he decided to place a vial of silver chloride solution outside of the blue-colored light from the prism. The solution outside of the blue end of the spectrum reacted even faster than any of the samples within the visible area.

- This experiment is remarkable because it clearly demonstrated the existence of light with a frequency even higher than the violet light at the edge of the visible spectrum—one that was not detectable to the human eye but had even more energy than visible light. Ritter had discovered ultraviolet light.

- Of course, since Ritter's day, the science of spectroscopy has come a long way. We no longer have to move a sample from one band of light to the next to observe how a sample is interacting with light. Instead, we rely on sophisticated instruments called spectrometers.

- All spectrometers have the same basic functional parts. How each part functions can be very different depending on the design and the technique needed to solve a particular problem, but all spectrometers consist of several crucial parts.

- First is a light source that generates a broad spectrum of light. This can be as simple as an ordinary incandescent lightbulb or a more advanced lamp. The light from this source is sent to a device called a monochromator. The simplest example of this would be Newton's prism dispersing sunlight into its constituent parts. Modern monochromators are usually more sophisticated and vary widely in design.

- The most common tool for this purpose is known as a diffraction grating, which accomplishes the same effect as the prism but can be very precisely machined to produce much narrower bands of very specific wavelengths, allowing us to make more detailed measurements.

- The light produced by the lamp and monochromator is passed through a sample one wavelength at a time. The sample can be solid, liquid, gas, or solution. Most commonly in the organic chemistry lab, we want to characterize the interaction of light with a compound in solution, so we use a quartz cell sometimes called a cuvette to hold the sample.

- The sample is held securely in the sample compartment, where light from the monochromator passes through it. As light exits the sample compartment, it strikes a detector that turns the light intensity into an electrical signal that can be interpreted by a computer.

- As the monochromator is adjusted to send different wavelengths of light to the sample, the detector measures the intensity of the exiting light at each wavelength and compares it to the intensity of the original beam when no sample is present.

- One of the first of these devices was built by German physicist August Beer, who is most well known for his law of absorbance, which he proposed in 1852. His law is concerned not with the maximum wavelength of absorption, but with just how strongly a given compound or sample absorbs.

- Beer observed that incrementally increasing the concentration of a chromophore, or colored chemical compound, cut the transmittance by half each time. Similarly, for a given concentration, the transmittance was cut in half when the length of the cell was increased incrementally.

- This relationship produces an exponential relationship. But scientists prefer to work with linear relationships, which are much easier to predict. So, Beer converted transmittance to a new unit, called absorbance. The absorbance is the negative logarithm of a sample's transmittance. The resulting plot is a very simple, very useful relationship in which incremental concentration or path length increases lead to incremental absorbance values.

- Beer's equation relates absorption to three important factors: a property of a given molecule called its molar absorptivity (which is essentially a measure of how well it absorbs at a given wavelength), the concentration of the sample, and the path length of the cell used in the experiment.

Factors Influencing Absorption: Pi-to-Pi* Transitions
- With conjugated alkenes, the frontier molecular orbital energy of conjugated compounds decreases as additional p atomic orbitals join the system.

- A single photon of light carries a specific packet of energy with it, and this energy can be calculated from its wavelength and Planck's constant using the relationship $E = hc/L$. Let's set the rule that this photon will only be absorbed by a molecule when it can promote an electron to a higher energy state when the orbital energy gap is exactly equal to the energy packet of the photon.

- Simple conjugated alkenes have pi systems that overlap very well with one another, creating a virtual superhighway for electrons to move from one orbital to the next. When electrons move from the highest occupied pi molecular orbital to the lowest unfilled pi molecular orbital, we call this transition the pi-to-pi* transition.

- Because of the fantastic geometric overlap of the associated orbits, this absorption occurs nearly every time an electron of sufficient energy strikes the molecule. In other words, the extinction coefficient for this absorption can be very, very large.

- We can construct an energy diagram for the molecular orbitals of a conjugated system, and when we do this for 1,3-butadiene and 1,3,5-hexatriene, we discover that the frontier molecular orbitals are closer in energy in the more conjugated compound. So, there is a distinct correlation between the wavelength of maximum absorption and the extent of conjugation in organic compounds. And this trend continues into the visible spectrum as well.

Ultraviolet Radiation Protection
- The single most well-known source of electromagnetic radiation is the Sun. In addition to the intense visible light emitted by the Sun, it is also constantly bathing the planet in strong ultraviolet radiation that is capable of causing damage to cellular DNA. The overall ultraviolet region of the spectrum is defined as the region between 100 and 400 nanometers (nm) in wavelength, but spectroscopists and other scientists break down the ultraviolet spectrum into three general regions: UVA, UVB, and UVC.

© amoklv/iStock/Thinkstock.

The single most well-known source of electromagnetic radiation is our Sun.

- UVC radiation spans 100 nm to 280 nm wavelengths. UVC is the portion of the ultraviolet spectrum that isn't normally found at the surface of the Earth, because it is naturally absorbed by the gasses in the Earth's atmosphere. This high-energy, potentially damaging radiation is absorbed by two naturally occurring atmospheric gasses in particular: oxygen, which absorbs between 100 nm and 200 nm, and ozone, which absorbs between about 200 nm and 280 nm. These two gasses work in concert to protect us from all of the most dangerous ultraviolet light reaching the planet.

- UVB radiation spans 280 nm to 315 nm wavelengths. It is the portion of the ultraviolet spectrum that reaches the surface and causes reddening of the skin, known commonly as sunburn. Because of this, it was long thought that UVB rays were the only dangerous form of ultraviolet radiation reaching the planet's surface, leading companies to produce sunblock formulations containing compounds like para-aminobenzoic acid (PABA), which absorbs very well right up to the edge of the UVB.

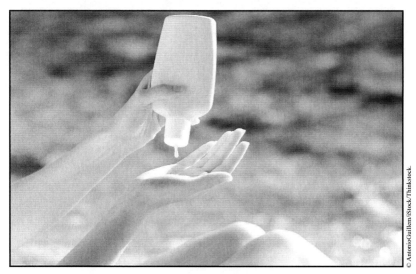

© AntonioGuillem/iStock/Thinkstock.

Modern sunblock formulations are effective against both UVA and UVB rays.

- UVA radiation spans the 315 nm to 400 nm region of the spectrum. For a long time, it was thought that this lower-energy ultraviolet light was not harmful to humans because it did not cause sunburn like UVB. In the later decades of the 1900s, however, medical researchers began to recognize that UVA light can, in fact, cause cellular damage even though it does not cause a reddening of the skin. So, modern sunblock formulations have abandoned PABA in favor of compounds like octocrylene and avobenzone.

- When we look at the ultraviolet spectrum as a whole and at those materials that we use to protect ourselves from it, both man-made and naturally occurring, the trend is that increasing the size of the conjugated pi system in the compounds correlates directly with the wavelength of maximum absorption in the compound. From oxygen, to ozone, to PABA, octocrylene, and avobenzone, each is larger and more conjugated than the next.

- So, using this predictive tool of extended conjugation, these products have been carefully engineered to have very large extinction coefficients at very specific wavelengths to fill in the gaps that Mother Nature left behind.

Suggested Reading

McMurry, *Fundamentals of Organic Chemistry*, 13.5, 13.6.

Wade, *Organic Chemistry*, 15.13.

1. Phenolphthalein is a compound often used to detect changes in pH because it changes from colorless to dark purple depending on pH. Under which conditions (low pH or high pH) do you expect phenolphthalein to be colored?

low pH   high pH

2. Although they usually represent the strongest absorption, the frontier molecular orbitals are not the only pi molecular orbitals that can participate in absorption. Do you expect other transitions within the pi system to absorb longer or shorter wavelengths?

UV-Visible Spectroscopy
Lecture 27—Transcript

In 1666, a young Isaac Newton first documented the separation of white light into its constituent visible colors using a prism. And in his notes on this now-famous experiment he described the rainbow of colors he saw as a spectrum, from the Latin word meaning apparition. What we didn't know at that time was that what he saw was only part of the story.

It was another 100 years, more than that, in 1801, when a young self-taught scientist by the name of Wilhelm Ritter acted on a hunch that maybe there was more to the spectrum than what Newton had seen with his own eyes more than 100 years earlier. The problem is that Wilhelm, he's a human, and his eyes couldn't see any further along the spectrum than Newton's could. So Ritter needed a different kind of eye, and he had just a tool for the job, silver chloride. Now, silver chloride is an inorganic compound, which is most famous for its use in photographic films. Solutions of silver chloride undergo a chemical reaction in the presence of visible light, which causes them to dissociate into a suspension of silver metal in chlorine gas. In other words, the solution turns dark when exposed to light.

It was known even before Ritter that this reaction did not proceed at the same speed in different colors of light. He could disperse light with the prism, then place vials of silver chloride solution in different parts of the resulting spectrum and clearly see that the reaction was fastest in blue, slower in green, and slower still in the red portions of the spectrum. Ritter's real contribution was that he decided to place a vial of silver chloride solution outside of the blue-colored light from the prism. His hunch paid off. The solution outside of the blue end of the spectrum reacted even faster than any of the samples in the visible area. Now this experiment is remarkable because it clearly demonstrated the existence of light with a frequency even higher than the violet light at the edge of the visible spectrum, one which was not detectable to human eye, but had even more energy than visible light. Ritter had discovered ultraviolet light.

If you're watching these lectures in order, I'd like to congratulate you on completing the structures and reactions segment of the course. You now have the basic knowledge and skills needed to understand and predict the reactivity of organic compounds on paper. But you must have asked yourself more than once during these lectures: How do we know how these molecules are reacting and behaving? How can these structures and the changes we make them be confirmed experimentally? After all, organic chemistry is about science, and science is about what we can prove, so where's the proof?

Well today we begin answering these very fair and very crucial questions by discussing the techniques which organic chemists and many others use to identify the products of their work. Today, we begin to explore the science of spectroscopy. Now, most simply put, spectroscopy is the observation of the interaction of light with matter. Before we can dive into spectroscopic techniques, we need to take a step back and characterize what we generally refer to as "light." Light is, in fact, electromagnetic radiation, a form of self-propagating energy composed of perpendicularly oscillating electric and magnetic fields. It exhibits both wave- and particle-like properties, and it carries energy and momentum.

Now, it's the wave properties of light and the energy it carries which are most interesting to spectroscopists, and we're going to tend to classify light based on those properties. Now, one of these properties associated with waves is the wavelength, or the distance from one point on the wave to the next analogous point. For example, the distance from the top of the peak on this end to the top of the next peak here on my wave, this is defined as the wave. Now, another property of light which interests us is its velocity. See, light has a fixed speed at which it travels in a given medium, regardless of its wavelength. Light in a vacuum always travels at 3 times 10 to the eighth meters per second.

Now, I'm going to start my wave in motion, but I'm not going to start it at the speed of light. So let's just move our wave now and pretend that this is the speed of light. Now, the frequency of light is the number of equivalent points along that wave which pass by a given location in a specified time. For example, my wave here has a frequency of about one oscillation per second. Now, since the maxima passed designated points in one-second intervals, that's my frequency. It sounds a little bit awkward to say this wave has a frequency of one per second, so we instead use a unit called hertz. This is a unit named in honor of German physicist Heinrich Hertz, who first characterized electromagnetic radiation in 1887. So, the constant speed of light becomes a proportionality constant for the relationship between wavelength and the reciprocal of frequency. For example, if I were to decrease my wavelength by a factor of two, I just increase the frequency by a factor of two.

One last property of light which is going to be of crucial importance to understanding spectroscopy, is this; light does still have some particle character. We call these particles of light "photons." A single photon of light contains an amount of energy which is directly related to its frequency by a factor called Planck's constant, sometimes shown in mathematical equations is the letter H. So, a single photon of light has an associated wavelength, frequency, and energy, all of which are related to one another mathematically.

So we classify light according to its frequency or wavelengths, which are related to the amount of energy carried by a single photon that light. We call this range of wavelengths the electromagnetic spectrum. The spectrum includes, in order of increasing frequency, and therefore, energy, radio waves, microwaves, infrared, visible, ultraviolet, X-ray, and gamma rays.

Radio waves can have wavelengths kilometers wide, and as we reduce the wavelengths to about the diameter of a baseball, we cross into the microwave region. Continuing to reduce the wavelengths to about the size of a human cell gets us to the infrared. The visible has wavelengths of several hundred nanometers, or about the size of a typical virus. And the ultraviolet begins to approach the size of some of the larger molecules that we've studied in the course, like proteins. Finally, from about 10 nanometers on down, x-rays constitute the smaller extremes of wavelengths, including the very high-energy gamma rays that have wavelengths smaller than the diameter of a single atom. So the electromagnetic spectrum spans a dramatic scale of wavelengths, and therefore, a dramatic range of energies per photon. Because of this, each class of radiation tends to interact with molecules in its own unique way, and scientists have devised a way to exploit all of them. So let's try spectroscopy experiment. You ready?

Here I have two glasses of water and flashlight. I'm going to shine my flashlight through the water. You can see that both glasses of water basically allow all of the visible light to transmit through them. Simple enough, right? Now I'm going to add just a small bit of red food coloring to one of the two glasses of water. I'll use my highly specialized scientific tool to stir that, to be sure it's well mixed. Now, I'll pass my white light back through the same two glasses, the original and the one with the added food coloring. You notice a difference? You see something different, don't you? Of course, the light reaching your eyes appeared white when it went through the first class because it contained only water, which didn't absorb anything, and you were able to see all the wavelengths of light which were generated by the flashlight. But in the second glass you saw the light that was passing through appear to be red, and that's because the contents of the solution were absorbing the complementary color of red, or green. So you know that something in that dye has to have the ability to absorb light of about 520 nanometer wavelengths.

Congratulations, you are now a spectra photometer; you just observed the interaction of light with matter, and measurements of the interaction of light with matter is the function of all optical spectroscopy instruments. Now what distinguishes optical spectrophotometers a lab from your eyes and mine is their increased sensitivity and the ability to see the electromagnetic spectrum with far greater precision over a far wider range of wavelengths than the human eye ever could. Instead of seeing a combination of all visible wavelengths simultaneously, spectrophotometers can give us information about each wavelength individually, telling us exactly which photons are absorbed by our sample and which photons are not absorbed.

Over the next few lectures, we'll take a look at some of the ways scientists coax structural information from molecules using spectrophotometers to sense how they interact with different forms of light. And we'll begin in the visible and ultraviolet regions of the spectrum today. Of course, since Ritter's day the science of spectroscopy has come a long way. We no longer have to move a sample from one band of light to the next to observe how a sample is interacting with light. Instead, we rely on sophisticated instruments called spectrophotometers to get this job done.

Now, all spectrophotometers have the same basic functional parts. How each part functions can be very different depending on the design and the technique needed to solve a particular problem. But all spectrometers or spectrophotometers, consists of several crucial parts. First, there's a light source, which generates a broad spectrum of light. Now, this can be as simple as an ordinary incandescent light bulb or a much more advanced lamp capable of producing ultraviolet. The light from the source is sent to a device called a monochromator. The simplest example of this would be one of Newton's prisms in between two slits. The purpose of the first slit is to ensure that all the light from the source reaches the monochromator in a parallel pathway.

Once it gets into the center the monochromator, we will see this beam being fanned out into its spectrum, just as Newton did in the 1600s. The final slit is there to allow only one very small region of that spectrum to pass on through.

The next element of our spectrophotometer is a beam splitter, which is really just a fancy array of mirrors which accomplishes the task of dividing our monochromatic light beam into two different beams which pass through two different regions of space. Now, these different regions of space are our sample and reference compartments. So the beam, which passes through my reference compartment, will encounter, really, nothing, just the reference. So this would be like a glass of water in the original experiment today. But the other beam passes through a sample cell, and that sample cell contains not just, say, water, but water with dye in it, as is our example before. So the difference between these two cells is the presence of a sample that we want to study.

Finally, the exiting light beams encounter detectors, and detectors have many different designs as well. But essentially what a detector does is it turns the impact of photons into some kind of an electrical signal which can be sent off to a computer which monitors the experiment. And it's the differences in the signals between the light exiting the reference cell and the light exiting the sample cell which are interpreted as the result of our experiment. So the advent of the modern spectrophotometer allows me to not only say that my sample appears red and therefore is absorbing green light. I can go much farther and determine exactly how much of every wavelength in the visible spectrum is being absorbed to create the color that I see.

But there's even more information available from this kind of experiment. One of the very first of these devices was built by a German physicist by the name of August Beer. Beer is most well known for his law the absorbence, which he proposed in 1852. His law as concerned not with the maximum wavelengths of absorption, but with just how strongly a given compound or sample absorbs. What Beer observed is that incrementally increasing the concentration of a chromophore, or light-absorbing molecule, cut the transmittance of a sample by one half each time he increased it. Similarly, for a given concentration of a sample the transmittance was cut in half when the length of the cell was increased incrementally.

Now this relationship produces an exponential relationship like the one I'm showing you here. But scientists prefer to work with linear relationships whenever possible because they're much easier to predict. So Beer converted transmittance into a new unit called absorbence. The absorbence is calculated by taking the negative logarithm of a sample's transmittance, and when this is done, the resulting plot is a very simple and useful relationship in which incremental concentration or path-length increase leads to incremental absorbence values as well.

So, Beer's equation relates absorption to three important factors, a property of a given molecule, called its molar absorb activity, which is essentially a measure of how well it absorbs at a given wavelength; the concentration of the sample; and the path length of the cell used in the experiment. So, if you know two of these three parameters for a given sample, measuring the absorbance would allow you to calculate the third variable using Beer's law. So this gives us a simple way to determine the concentration of solutions of known compounds, or, to help identify compounds in solutions of known concentrations.

Now we saw on our discussion of conjugated alkenes how the frontier molecular orbital energy of conjugated compounds decreases as additional p-atomic orbitals join the system. So, let's begin by restating a crucial proviso here, that a single photon of light carries a specific packet of energy with it, and that this energy can be calculated from its wave length and from Planck's constant using a relationship of, energy is equal to Planck's constant times the speed of light, divided by the wavelength of that particular photon.

So for example, using this equation I can determine that a photon with a 520-nanometer wavelength carries with it exactly 8.83 times 10 to the minus 19th joules of energy. No more, no less. We sometimes convert these energies from joules per photon into kilojoules per mole of photons to make the numbers a little more manageable. So in our example, 3.82 times 10 to the 19th joule per photon becomes 230 kilojoules per mole of photons. Second, let's set the rule that this photon will only be absorbed by molecule when it can promote an electron to a higher energy state. In other words, when the orbital energy gap between occupied and unoccupied molecular orbitals are exactly equal to the energy packets of photons.

Now let's take a look at a few molecules with distinctive UV spectra, compounds with which we're pretty familiar by now, the conjugated alkenes. Simple conjugated alkenes have pi systems which overlap very well with one another, creating a virtual superhighway for electrons to move from one orbital to the next. When electrons move from the highest occupied pi molecular orbital to the lowest unoccupied pi molecular orbital we call the transition a pi-to-pi star transition. Now, because of the fantastic geometric overlap of the associated orbitals, this absorption occurs nearly every time an electron of sufficient energy strikes that molecule. In other words, the extinction coefficient for the absorption can be very, very large.

Now recall from our lecture on conjugated alkenes that I can construct an energy diagram for the molecular orbitals of a conjugated system, and that when we did this for 1,3 butadiene and 1,3,5 hexatriene, we discovered that the frontier molecular orbitals are closer in energy in the more conjugated compound.

Now, if we were to carefully calculate the energies of the frontier molecular orbital gaps, we would see the following trend for the first three alkenes in the series: 686 kilojoules for ethane; 540 kilojoules for 1,3 butadiene; and 452 kilojoules for 1,3,5 hexatriene. Assuming that our earlier rules hold, this means that promoting one electron from the homo of each compound to the lumo should require a photon with the exact same energy. So ethane should require a photon of 686 kilojoules per mole, or, one with the wavelength of 171 nanometers; 1,3 butadiene requires a photon of 217 nanometers; and 1,3,5 hexatriene requires a photon of 258 nanometers.

So there is a distinct correlation between the wavelength of maximum absorption and the extent of conjugation in organic compounds. And this trend continues into the visible spectrum as well, where compounds, like beta carotene, with its 11 conjugated double bonds, has a maximum absorption at 454 nanometers.

So let's take a look at another example in the form of a compound so well understood that it's frequently used a reference standard for experiments like these. I'm talking about our old friend benzene. Benzene has a maximum absorption at 220 nanometers, lower than 1,3,5 hexatriene, because promoting electrons causes benzene to lose some of the aromaticity. But let's take a look at the maximum absorption of benzene's cousins naphthalene and anthracene. Now see, when we do, we notice that the overall trend of increasing lambda max, or, maximum wavelength of absorption, with increasing pi-system size holds for our aromatics as well.

Of course, no discussion of the ultraviolet region of the spectrum is ever complete without the mention of the single, most well-known source of electromagnetic radiation, the Sun. In addition to the intense visible light emitted by the Sun, it also constantly bathes the planet in strong ultraviolet radiation, which is capable of causing damage to cellular DNA. The overall ultraviolet region of the spectrum is defined as a region between 100 and 400 nanometers in wavelength, but spectroscopists and other scientists break the UV spectrum down into three general regions, which are commonly called UVA, UVB, and UVC. So let's look at those one by one.

UVC radiation spans 100 nanometers to 280 nanometers in wavelength. UVC is that portion of the UV spectrum which isn't normally found at the surface of the Earth's naturally absorbed by the gases in the Earth's atmosphere. This high-energy, potentially damaging radiation is absorbed by two naturally occurring atmospheric gases in particular, oxygen, which absorbs between 100 and 200 nanometers, and ozone, which absorbs between about 202 and

280 nanometers. So these two gases work in concert to protect us from all of the most dangerous UV light reaching the planet.

UVB radiation spans to 280 to 315 nanometer wavelengths. It's that portion of the UV spectrum which reaches the surface and causes reddening of the skin which we commonly refer to as sunburn. Because of this, it was long thought that UVB rays were the only dangerous form of UV radiation reaching the planet's surface, leading companies to produce sunblock formulations containing compounds like para-aminobenzoic acid, or PABA, which absorbs very well right up to the edge of the UVB.

UVA radiation spans the 315 to 400 nanometer region of the spectrum, and for a long time it was thought that this lower-energy UV light wasn't harmful to humans because it didn't cause sunburn the way that UVB does. In the later decades of the 1900s, though, medical researchers began to recognize that UVA light can, in fact, cause cellular damage, even though it doesn't cause a reddening of the skin. So modern sunblock formulations have abandoned PABA in favor of compounds like octocrylene and avobenzone, shown here as its enol tautomer.

So, when we take a look at the UV spectrum as a whole, and all those materials which we use to protect ourselves from it, both man made and naturally occurring, we should start to notice a distinct trend. Increasing the size of the conjugated pi system in the compound correlates directly with their wavelength of maximum absorption, from oxygen, to ozone, to PABA, octocrylene, and avobenzone, each larger and more conjugated than the next. So, using this predictive tool of extended conjugation, these products have been carefully engineered to have very large extinction coefficients at very specific wavelengths to fill in the gaps that Mother Nature left behind.

Some aromatic and conjugated compounds are also titratable, meaning that they can accept and release protons, which can change their absorptive properties. So let's belly up to the bar here. Take for example the anthocyanins in this red wine, well, blush wine. Now, anthocyanins are polyphenols, and that means that not only are they highly conjugated, which gives them beautiful colors, but they also have hydroxyl groups on them, which can be protonated and de-protonated. Now, if you're an oenophile, you might want to look away for this demonstration, because I'm going to take this wine, which is currently the pH of about 5, and I'm going to bring it up and we're going to watch the effect it has on the color.

Now, to do this, I'm going to use a solution of sodium hydroxide, which is caustic, so I'm going to find a glove that isn't destroyed, and I'm going to cover up my hands just for safety's sake. Of course, what I'm about to do is also going to render the wine undrinkable, so I'm probably going to make a few enemies or at least not make any friends by doing this today. So here we have the original one; you can see it's clearly red in color. This is because of a compound called malvadin glucoside. And I have here a small Erlenmeyer flask with an aqueous solution, that's water-based, of sodium hydroxide.

Now, as a control, just to prove to you that there's nothing magical about what I'll be adding to these wines, I'm going to put a few drops of my solution into a glass full of water. And of course, we see no color change or reaction of any kind taking place. So, clearly, if we see any changes in these wines as a result of adding my hydroxide to it, it has to be coming from something already in the wine.

I'll leave my second glass unaltered, as our control, so we have a reference. And now I'm going to add a little bit of my aqueous sodium hydroxide to my first victim today. Did you see a change? I know I certainly do. The change that's taking place here is a result of deprotonating some of those phenolic compounds, most notably, the malvadin glucoside in the red wine.

But I can go in farther than that and add a very large amount of my sodium hydroxide to my third glass. Let's give a little bit more just for good measure. There we go. All right. I'll cap my sodium hydroxide solution and give my wines one lasts swirl to be sure that they're thoroughly mixed. Now, what should be very apparent here is that something has changed. Clearly, changing the pH conditions of the wine has altered the structure of these compounds. It's quite an impressive color change, isn't it?

So let's recap what just happened here. The compounds which impart the color to the wine, like malvadin glucoside, have become deprotonated, and removal of one or more hydroxyl protons allows the newly deprotonated oxygen atoms to partake in resonance with the rest of the aromatic system, thereby extending the conjugation and changing the frontier molecular orbital energies of the color-giving organic species in the wine.

I've actually prepared some more samples here of the same wines that have been altered with sodium hydroxide as well in varying proportions, and you can see that there's actually a broad range of colors that can be achieved in this way.

Now, as the conjugated pi systems here grew in energy, the pigments absorbed first in the green; then in the orange, which resulted in a blue color; and finally in the red portion of the spectrum which your eyes detect as green. And that is the chemistry of destroying a perfectly good rosé.

So today we started on our journey through the task of probing molecular structures, We talked a little bit about light in the electromagnetic spectrum which comprises it. We explored the relationships among wavelength, frequency, and energy in any given photon, and we noted that a single photon carries a packet of energy which can only be absorbed in a transition of equal energy.

We talked about UV light, in particular, how Wilhelm Ritter discovered this type of electromagnetic radiation using decomposition of silver chloride solutions with light separated by a prism. We examined the general structure of a UV-visible spectrophotometer, from light source, to monochromator, to beam splitter, sample-in reference compartments, and finally, detectors, and how these units all work in concert to produce a UV-visible spectrum.

Next we discussed how extending conjugation in a compound leads to lower energy HOMO-LUMO transitions within the pi system of molecules, lengthening the wave lengths of the associated molecules' maximum absorption. We saw how this concept can be used to explain how the Earth's atmosphere protects us from UVC radiation, but we need to devise our own methods to protect ourselves from UVB and UVA radiation.

And finally, you witnessed me doing the unthinkable and rendering a perfectly good wine undrinkable by altering its pH. In doing so, however,

I demonstrated that the protonation state of many organic compounds can profoundly affect their absorption characteristics as well.

Next time, we're going to turn our attention to the other side of the visible spectrum and explore its other neighbor, infrared light, and how it interacts with molecules by a completely different mechanism, yielding completely different information about structures. And I promise I won't destroy one single beverage during the entire lecture. I'll see you then.

Infrared Spectroscopy
Lecture 28

J ust like ultraviolet light, infrared light from the Sun is constantly bombarding us. Unlike ultraviolet light, however, infrared light isn't associated with any particularly dangerous health effects. Infrared is still a very interesting part of the electromagnetic spectrum, though. It interacts with organic molecules by a completely different mechanism, yielding completely different information about their structure. In this lecture, you will learn about the red end of the visible spectrum and what lies beyond.

Infrared Light

- Objects at or near room temperature naturally emit the wavelength of light known as infrared. Scientists like those at NASA have developed cameras and telescopes capable of detecting the infrared emissions of both small objects on Earth and celestial bodies on the other side of the known universe, creating false-color images, in which the detected infrared light has been shown.

Though invisible to the naked eye, infrared light can be captured using sensitive cameras.

- But infrared light isn't just radiated. Like all other parts of the electromagnetic spectrum, it can also be absorbed. And just as a warm object cools by radiating energy as infrared light, an object can be warmed by absorbing it.

- Wilhelm Ritter's experiment that resulted from his hunch to look outside of the blue end of the visible spectrum was almost certainly inspired by one conducted earlier that year by Sir William Herschel, who made a truly inspired discovery when he asked a simple question: Which color of light is most effective at heating a surface on which it is shined? To answer this question, he used a prism to diffract sunlight into a spectrum and placed thermometers with blackened bulbs at regular intervals within that spectrum.

- Herschel saw that the temperature indicated by the thermometers increased across the spectrum from blue to red. Knowing what we do about the amount of energy contained in photons of different wavelengths, this may seem counterintuitive at first, but Herschel essentially proved that red light is absorbed more efficiently by objects near room temperature.

- But the pertinent result of Herschel's experiment wasn't the trend he saw in the visible spectrum. His real breakthrough came when he wondered if the trend continued beyond the spectrum that he could see. Sure enough, when he placed a thermometer just beyond the red portion of the spectrum, it read an even higher temperature than the one in the red light. Clearly, there was some form of light beyond the red portion of the spectrum, prompting him to call his new discovery "infrared" light.

Infrared Spectrometers
- Modern spectroscopy in the infrared region is conducted using an instrument very similar to those used for UV-visible. Both start with a source lamp, though the infrared source lamp is made of a different material that generates more infrared light.

- The infrared beam is split using a mirror, which sends the beam to two different cells. One is the reference cell, containing no sample, and the other is the sample cell, containing the molecules we want to analyze.

- The beams are then reflected back onto the same path using a device called a beam splitter, which allows only one of the two beams to make it to the detector at a given time. After passing through a monochromator, the intensities of the two beams are compared at the detector to calculate the transmittance of a sample.

Vibrational Modes
- In order to understand infrared spectroscopy, we need to begin to think of molecules in terms of dynamic models, rather than the idealized structures—symmetrical, rigid, unmoving—that are so often depicted in the literature, models, textbooks, and television. Dynamic simulations approximate the random motions of each atom within the molecule. The bonds between and among the atoms stretch, bend, and twist in numerous ways. Put simply, bonds vibrate.

- Consider carbon monoxide—two atoms joined by a triple bond. With only two atoms, there is only one possible bond vibration that can take place: a stretching between the atoms. But when we move on to carbon dioxide, a molecule with three atoms, we can conceive of vibrations involving a symmetrical stretch, an asymmetrical stretch, symmetrical bending, and asymmetrical bending. So, there are four different ways in which CO_2 can vibrate.

- If we continue adding atoms, we find that the total number of vibrational modes possible is equal to $3n - 5$, where n is the number of atoms in a linear molecule—or $3n - 6$ for nonlinear molecules. This means that for a compound as simple as methane, we expect to find 9 different types of vibrations. Ethane has 24, and more complex molecules have dozens or even hundreds of vibrational modes, each with its own frequency.

Infrared Absorption
- What is it that brings these two apparently disparate concepts—how infrared spectroscopy is conducted and the vibration of bonds and sets of bonds in organic molecules—together? The answer is

that many bond vibrational frequencies are the same as those of infrared light, and just as an electronic transition could be promoted by just the right photon in UV-visible spectroscopy, a transition in the vibrations of a molecule can be induced by just the right infrared photon.

- When a bond vibrating at a particular frequency encounters a photon of identical frequency, that photon can be absorbed, increasing the amplitude of the associated vibration. This is the absorption that is measured by the spectrometer. For example, if our spectrometer reports to us that photons with a frequency of 2000 hertz are being absorbed, we know that there must be a bond vibrating at that frequency within the molecule.

- Of course, we have a pretty good idea of the vibrational frequencies of different types of bonds near room temperature. The trend is that larger atoms and weaker bonds lead to slower vibrations. In other words, a C-H bond should vibrate faster than a carbonyl bond, which should vibrate faster than an ether bond. In fact, this is exactly what we see in infrared spectroscopy.

Detecting Functional Groups with Infrared
- There is one more catch to the science of infrared spectroscopy. It is known as the infrared selection rule, which states that a vibrating bond or set of bonds can only absorb a photon of the same frequency if that vibration causes a change in the dipole of the molecule. No dipole change means no absorption; strong dipole change means strong absorption.

- Functional groups, which usually have strong dipoles, are often powerful infrared absorbers, while less-reactive bonds, such as alkyl C-H bonds, tend to absorb very weakly or sometimes not at all.

- Most functional groups contain good, strong polar bonds to at least one hydrogen or double and triple bonds among larger atoms (or both). So, the signatures associated with most common functional groups in infrared are typically found in the higher-frequency region of the spectrum, whereas the vibrations within the skeleton have a tendency to cause absorptions in the lower-frequency region of the spectrum.

- The few but distinct absorptions caused by functional groups lead us to define the higher frequency region of the infrared spectrum as the functional group region, because it is often where we find the information we need to catalog which functional groups are present in the sample and which are not.

The Fingerprint Region of the Infrared

- The high-frequency region of the infrared spectrum tells us about the presence or absence of functional groups, but what about the low-frequency region? If it is so complex that it defies dissection, why do we bother to collect this information?

- We call the low-frequency region of the spectrum the fingerprint region, because just like fingerprints, the complex set of vibrations within the skeleton of a molecule give it a complex pattern in this region that is unique to that particular molecule. Just as a fingerprint can be compared to a reference to identify its owner, the fingerprint region of an infrared spectrum can be compared to a standard to identify a molecule.

Suggested Reading

McMurry, *Fundamentals of Organic Chemistry*, 13.3, 13.4.

Wade, *Organic Chemistry*, 12.1–12.2.

1. Infrared spectroscopy samples must be kept exceedingly dry to produce the best spectra. Explain why.

2. Which bonds in hydrofluorocarbons (Lecture 10) give them the potential to be a greenhouse gas, trapping heat by infrared absorption?

Infrared Spectroscopy
Lecture 28—Transcript

Last time, we started considering how spectroscopists use light as a means of probing the structured and behavior of molecular systems. Specifically, we took a look at the ultraviolet and visible regions of the electromagnetic spectrum and how radiation of this energy is perfectly suited to excite electrons from one molecular orbital to another, giving us a way to probe the energetics of these kinds of systems.

Today we're going to take a look at the red end of the visible spectrum and what lies beyond that. Just like UV light, infrared light from the sun is constantly bombarding us. Unlike UV light, though, infrared light isn't associated with any particularly dangerous health effects. Infrared is still a very interesting part of the electromagnetic spectrum, though. Objects at or near room temperature naturally emit this wavelength of light, and that includes you. Believe it or not, you glow in the dark.

The catch is that you glow in the dark at too long of a wavelength to be detected by the human eye. Scientists, like those at NASA, have developed cameras and telescopes capable of detecting the infrared emissions of both small objects here on Earth and of celestial bodies on the other side of the known universe, creating false color images like these in which the detected infrared light has been shown as different colors. But infrared light isn't just radiated. Like all other parts of the electromagnetic spectrum, it can also be absorbed. And just as a warm object cools by radiating energy as IR light, so can an object be warmed by absorbing it.

Now, last time I made it sound like Ritter was stricken with a divine inspiration to look outside the blue end of the visible spectrum. But in fact, his experiment was almost certainly inspired by one conducted earlier that year by another English scientist, Sir William Herschel. Herschel was a prolific astronomer, a composer, and a mathematician, a true Renaissance man. And he was the one who made a truly inspired discovery when he asked the simple question: Which color of light is most effective at heating a surface when shined on it? To answer this question he used a prism to diffract sunlight into a spectrum and placed thermometers with blackened bulbs at regular intervals within that spectrum.

So let's bring up this IR spectrum, or, visible spectrum, with infrared just outside of it. So, Herschel saw that the temperature indicated by his thermometers increased across the spectrum from blue to red. And knowing what we do about the amount of energy contained in photons of different wavelengths, it might seem a little bit counterintuitive at first. But Herschel essentially proved that red light is absorbed more efficiently by objects that are near room temperature.

But the pertinent result of Herschel's experiment wasn't the trend he saw in the visible spectrum. His real breakthrough came when he wondered if the trend continued beyond the spectrum that he could see. And sure enough, when he placed another thermometer just beyond the red portion of the spectrum, it read an even higher temperature than the one in the red light. Clearly, there was some form of light beyond the red portion of the spectrum prompting him to call his new discovery, infrared light. Today, we're going to take a look at how modern scientists use Herschel's discovery to probe the structures of organic compounds and make new discoveries today, more than 200 years after its initial discovery.

Modern spectroscopy in the infrared region is conducted using instruments very similar to those used for the UV visible. Now, the simplest of these will follow the same format, starting with the light source, but in this case, a light source which emits powerful infrared light. That light will, again, go through a monochromator, which will selectively direct one wavelength at a time through beam splitter. The beam splitter will split the beam into two, giving us two different beams of infrared light, one traversing a reference cell, one traversing a sample cell, which is exactly the same but contains one additive, and that's the molecules that we want to analyze. As the beams exit, their intensities are measured by two detectors and compared. Now there are much more sophisticated apparatus, but this is the simplest way to conduct, and therefore, to understand infrared.

OK, so, now we know how scientists shine infrared light on samples and measure which wavelengths of light are absorbed, but we still need to try to understand exactly why anyone would want to do such a thing. Now to explain this, I want to show you a molecule called fluorenone on the big screen here using my ChemDraw chem 3-D software. Now, we're used to seeing molecules in this kind of configuration when we see them in textbooks and papers. Notice that all of the bond lengths and angles are identical and ideal. The molecule appears perfectly planar. Every bond in my benzene rings is exactly 120 degrees. Everything looks exactly as it should, right? But do molecules really look like this? Do they really act like this? Are they rigid, unmoving, and symmetrical? No, of course not. This representation is done using average bond angles and average bond lengths of a dynamic set of bonds. They really look more like this.

What's running right now is a dynamic simulation which approximates the random motions of each atom within a molecule. The bonds between and among those atoms are stretching, bending, and twisting in many different ways. Just put more simply, bonds vibrate. They don't all vibrate in the same way, either. Notice that the carbon-oxygen bond is oscillating very slowly compared to some of those carbon-hydrogen bonds. And if you watch for a while, you can see not only individual bonds stretching and compressing, but other sets of bonds are opening and closing like scissors and even twisting. Now, this is really evident when I turn the molecule to look at it edge on. So instead of being perfectly flat and planar, you'll notice that some of the vibrations cause the molecule to flex and twist in and out of the plane that we usually use to represent it.

So let me stop my simulation now and bring my molecule back to its idealized state. So we're used to seeing the molecules drawn this way, but, to the eyes of a seasoned chemist, it's apparent that they really aren't like this. When it comes to actually modeling how molecules behave, we have to learn how to think of a picture like this in terms of that animation that we saw previously. Now, in order to understand infrared spectroscopy, we need to begin to think of molecules in terms of that dynamic model that I just showed you, rather than the rigid, idealized structures that are so often depicted.

So let's begin breaking this down by looking at a set of much simpler molecules and how they behave. First, let's consider carbon monoxide, two atoms joined by a triple bond. Now, with only two atoms, there's only one possible bond vibration which can take place, a stretching between the atoms. But what if we move on to carbon dioxide, a molecule with three atoms? Well, now we can conceive of vibrations involving a symmetrical stretch, an asymmetrical stretch, a symmetrical bend, and an asymmetrical bend. So, there are actually four different ways in which a CO_2 molecule can vibrate.

If we continue adding atoms, we find that the total number of these vibrations, which we call vibrational modes is equal to $3n - 5$, where n is the number of atoms in a linear molecule. Or, $3n - 6$ for non-linear molecules. So, this means that for a compound as simple as methane, we expect to find nine different types of vibrations going on within that molecule. Ethane has 24, and even more complex molecules like fluorenone known have dozens, or even hundreds of vibrational modes, each with their own distinct frequency.

So, we've now covered how IR spectroscopy is conducted and the vibration of bonds and sets of bonds in organic molecules. And I'm sure by now you're wondering, what is it that brings these two apparently disparate concepts together? The answer is this; many bond vibrational frequencies are the same as those of infrared light. And just as electronic transitions could be promoted by just the right photon in a UV visible experiment, so can transitions in the vibrations of a molecule be introduced by just the right infrared photon.

When a bond vibrating in a particular frequency encounters a photon of identical frequency, that photon can be absorbed, increasing the amplitude of the associated vibration in the process. This is the absorption which is measured by the spectrophotometer. Now, the frequencies which correspond to the vibrations of bonds in organic molecules are on the order of billions of cycles per second or billions of hertz, much faster than the model I showed you. This produces some pretty unwieldy numbers for frequency, of course, so in IR spectroscopy we instead use a unit called wavenumbers.

A wavenumber is simply the reciprocal of a photon's wavelength in centimeters. For example, light with a wavelength of five microns or 0.0005 centimeters, which is a typical IR wavelength, will have a wavenumber of

2,000—2,000 reciprocal centimeters. Of course, as you might imagine, we have a pretty good idea of the vibrational frequencies of different types of bonds near room temperatures. It's actually easier to predict than you might think; you simply have to think of atoms and bonds as though they were weights and springs.

Let me demonstrate this for you. If I connect a set of light weights by a fairly stiff spring and give it some energy, the vibrational frequency is expected to be fairly high. And this is because the light weights have only a small amount of momentum, which is easily overcome by the strong spring in order to reverse the mass's direction. Of course, right now, we have nothing to compare this to, so let's make a change and connect a heavier weight to the same kind of spring. You see that the spring needs more time to overcome the greater momentum of the larger weight, causing the frequency of the oscillation to be lower. I can also alter the frequency by adjusting the stiffness of the spring. For example, connecting similar weights to a stronger spring means that the spring can overcome the momentum of those weights more easily and reverse the motion of the mass more quickly. So we've increased the vibrational frequency in this case.

So, larger masses and weaker springs lead to slower vibrations. If we translate this phenomenon from the macroscopic example of weights and springs to the molecular level, atoms would be our weights, and the chemical bonds between them would be our springs. We would say that larger atoms in weaker bonds lead to slower vibrations. In other words, a CH bond should vibrate faster than a carbonyl bond, which should vibrate faster than a carbon-oxygen-ether bond. In fact, this is exactly what we see in IR spectroscopy.

Now, there's one more catch the science of IR spectroscopy; it's known as the IR selection rule. Now, the IR selection rule states simply this, that a vibrating bond or set of bonds can only absorb a photon of the same frequency if that vibration causes a change in the dipole of the molecule. No dipole change means no absorption. Strong dipole change means a strong absorption.

So, it should come as no surprise that functional groups, which usually have strong dipoles in their bonds, are often powerful IR absorbers. While, less reactive bonds, like alkyl CH bonds, tend to absorb very weakly or sometimes not even at all. Now you probably noticed that most functional groups contain good, strong polar bonds to at least one hydrogen, or double and triple bonds among larger atoms, or sometimes both. So the signatures associated with the most common functional groups in infrared are typically found at higher frequency regions of the spectrum, whereas, the vibrations within the skeleton tend to have absorptions in the lower frequency region of the spectrum on the other side.

The few but distinct absorptions caused by functional groups leads us to define the higher frequency region of the IR spectrum as the functional group region. Because it's often where we find the information we need to catalog which functional groups are present in the sample and which are not. But this region tells us very little about the carbon-carbon bonds within the scaffold of the molecule, which is usually fairly large with many complex vibrational modes, too many to consider individually. But these types of molecular vibrations tend to have lower-frequency IR absorptions, so we've labeled that region of the spectrum the fingerprint region for the fact its complexity makes it difficult interpret, but, unique to each molecule which generates it.

So, let's take a look at a few IR spectra and see if we can make sense of the information that they contain, starting with the functional group region of some common hydrocarbons. Let's take a look at three different hydrocarbons with which we're very familiar. The saturated hydrocarbon, pentane; 1-pentene, a terminal alkene; and 1-pentyne, a terminal alkyne. So these molecules are all very similar to one another structurally with one distinct difference, that difference being that the degree of saturation, or unsaturation, is changing.

So let's begin by taking a look at the infrared spectrum of pentane. Now, we've taken these spectra from a database called the AIST database. They're available online free of charge, so if you'd like to check them out for yourself, be sure that you find the URL and have a look. It's a very, very broad database with lots of really good data to see, certainly more than I'm going to show you today.

So let's start with pentane, again. Here's my structure, it's a simple hydrocarbon consisting only of carbon-carbon single bonds and carbon-hydrogen single bonds. So I expect to see primarily evidence of those kinds of bonds in a molecule. Now, looking at the functional group region, which is over here on the left-hand side at my higher frequencies, and therefore, higher wavenumbers, I see that there's a very strong absorbance at about 2,900 wavenumbers. And this is a classic signature for an sp^3-hybridized carbon-hydrogen bond. The rest of this is the fingerprint of the molecule, which corresponds to the backbone atoms and their particular vibrations. So this is a very simple infrared spectrum of a very simple hydrocarbon and a great place to start.

But when I move to 1-pentene, I'm going to introduce a few changes to the molecule. Most notably, the hybridization of two of the carbon atoms which bear CH bonds. By changing their hybridization state from an sp^3 to an sp^2, I'm going to increase the s-character of those bonds, and that means I'm going to make them a little bit tighter, little bit tougher, and that means, their frequency should increase ever so slightly. So let's keep our eyes on the functional group region as I switch my spectrum over from pentane to 1-pentene.

Here it is, a classic signature of an sp^2-CH bond. I now have an absorption not only at 2,900 wavenumbers corresponding to my sp^3-hybridized carbon-hydrogen bonds, but I also have a new absorption, which is taking place at about 3,100 wavenumbers. Now remember, my axis here is transmittance, so my peaks are actually valleys. So the lower the spectrum goes, the stronger the absorption is.

OK, so let's move on to 1-pentyne after we think for just a moment about what kind of change we expect to see in the functional group region. I'm going to be changing the hybridization state of at least two of my atoms here. These two we're going to go from sp^2 to sp^3. Excuse me, they're going to go from sp^2 to sp. And that means they'll have even more s-character, and having even more s-character means I expect to see, that's right, an infrared absorption at an even higher frequency. So let's make that change now. Sure enough, here is my absorption at about 3,300 wavenumbers, a perfect signature for an *sp*-hybridized carbon-hydrogen bond just like the one I have in 1-pentyne.

But remember, we talked about the importance of polarity and how IR is a great tool for probing the structures and presence of functional groups. So let's put some more functional groups in and see what happens to our IR spectra. Let's compare one pentanol to pentanoic acid, both compounds which we know have very polar CE and OH bonds. Well, the IR spectrum of 1-pentanol has a very clear signature for an OH; it's this large broad peak here. Now, the peak is fairly broad because OHs have a tendency to hydrogen bond to one another and exchange back and forth between and among molecules, and so this broadens their peaks out a little bit because their chemical environments can change as the experiment progresses. Nonetheless, I can see it very clearly here. And on top of that peak, I can also see my sp^3 carbon-hydrogens. So what I have here is a very obvious functional group region for an aliphatic alcohol.

Now, if I make a slight change here and oxidize my 1-pentanol down to pentanoic acid, let's what happens. OK, two changes that we can point to directly. The first of which is that the OH peak has gotten broader, and also that this very large absorbance has come into my spectrum at around 1,700 wavenumbers. So we can easily explain this broadness here by, again, the exchange of my OH bond from the pentanoic acid, not the bond itself, mind you, but the hydrogen being exchanged. But what's causing the really, really powerful absorbance here at 1,700, at a lower frequency? That's the presence of carbonyl. Heavier carbon and oxygen atoms lead to lower frequency vibrations in that particular bond. So once again, my functional group region has given me a very clear signal here. I have an aliphatic organic acid.

Let's continue and do a few more functional groups. How about the IR of ketones and esters, some more carbonyl-containing compounds? Now, we should be able to use what we know so far to get a rough idea of what these will look like; 2-pentanone has a very clean IR spectrum here. Obviously aliphatic, because we have strong absorptions here at about 2,900. Now, the strong absorption isn't because the carbon-hydrogen bond is very polar, actually; the strong absorption is because there are so many of those bonds in the molecule. But in addition to that, I have my carbonyl absorbing at about

1,700 wavenumbers. So this is a very obvious functional group region for an aliphatic ketone. It shares some of its features in common with my organic acids, but it's missing the critical OH signal which would tell me it's an acid. So, you need to think in terms of not only what's present in your spectrum when you're doing an analysis, but also what's not there. The absence of peaks can be just as informative as the presence.

Let's move on to butyl acetate, an ester compound. Very similar, isn't it, to 2-pentanone, sp^3 CHs, and my carbonyl. But there's another very strong absorbance which is working its way down into the fingerprint region, but it's so strong that I can easily pick it out. It's right here. And this is one of the absorbances which is commonly caused by the presence of a carbon-oxygen single bond. Remember, double bonds are stronger, so they create a higher frequency vibration. Carbon-oxygen single bonds would be weaker but have the same-sized atoms, therefore we see a lower frequency vibration.

One last look at the fingerprint region. Let's take a look at the IR of amines. I'm going to compare a primary, secondary, and tertiary amine here. First let's look at the primary amine, pentylamine. Again, easily seen as an aliphatic, and we also have two absorptions that are slightly higher in frequency than my aliphatic sp^3 CHs. And those aren't from an alkyne proton. Those are from the NH protons here. This is from an asymmetric stretch and symmetric stretch mode. In other words, this is the interpretive dance portion of the course now, in other words, if you think of my head is a nitrogen atom and my hands as being the hydrogens, there are two vibrational modes there, because two hydrogen atoms means they can vibrate symmetrically or asymmetrically, relative to the nitrogen, hence, two peaks in the NH region of my spectrum.

But what if, instead, I have a secondary amine, like N-methylbutylamine? Well, let's watch the spectrum change. A lot of similarity there in the functional group region, but one very crucial change. There's only one absorbance now in the NH region of the spectrum. So, this tells me that there's only a single NH, and therefore, only a single vibrational mode, only a single stretching mode, which can appear.

And the trend continues if I move onto a tertiary amine, like ethyldiisopropylamine. Let's take a look at the spectrum. Again, the functional group region looks very similar to the others, except that there are now no NH stretches to be found whatsoever. So, we cannot only detect the presence of amines, but we can detect whether they're primary, secondary, or tertiary, simply using the functional group region of the spectrum.

So the high frequency region of the IR spectrum tells us about the presence or absence of functional groups. But what about that lower frequency region? If it's so complex that it defies dissection, why do we bother to collect that information? Well, we call the low frequency region of the spectrum the fingerprint region, because just like fingerprints, the complex set of vibrations within the skeleton of a molecule give it a complex pattern in this region, which is unique to that particular molecule. So, just as a fingerprint can be compared to a reference to identify its owner, so can the fingerprint region of an IR spectrum be compared to a standard to identify a molecule.

Let's work through a quick example here to see if I can illustrate my point. I'm going to show you the IR spectra of four different aliphatic esters—butyl acetate, isobutyl acetate, *sec*-butyl acetate, and *t*-butyl acetate. All of these have the same molecular formula, and they all have the same kinds of bonds, carbon-carbon singles, carbon-hydrogen singles, carbonyls, and carbon-oxygen singles. So, when we look at the functional group region of these molecules and try to tell them apart, we're going to be a little disappointed. Let's start with butyl acetate. It can break down the spectrum very easily; 2,900 wavenumbers means its aliphatic, lots of carbon-hydrogen bonds with sp^3 carbons, clear presence of a carbonyl and also the clear presence of a CO single bond.

But what about all these other little absorptions in here, are they really useless? No, they're not. Because when I switch my spectrum from butyl acetate to isobutyl acetate, I want you to keep your eye on the functional group region the first time. Hardly anything has changed. So, looking at this from a functional-group-region perspective, the two spectra appear almost identical.

Let me go back again to butyl acetate. All the same maxima, all the same peaks in essentially the same frequencies. So, could we tell them apart from one another? And the answer, of course, is yes, as long as I have a reference spectrum. If I know the specific pattern that occurs in the functional group region, take a look at this now as I change the fingerprint region, excuse me, take a look at the fingerprint region as I switch the spectrum back to isobutyl acetate once more. We see some specific changes now. And if I switch to sec-butyl acetate, even greater difference. Moving over to *t*-butyl acetate, even more variability. So even though I can't dissect every peak there and break them all down into their individual vibrational modes. In the case of the fingerprint region of the spectrum, I don't have to. Because I can always go to a data library, just like the AIST, and compare it to the data that I see when I run my experiment.

So let's summarize what we've learned about FTIR spectroscopy today. We talked about what lies beyond the red portion of the visible spectrum, a form of radiation with lower energy than red light, earning it the name infrared. We took a brief look at the layout of a very basic IR spectrophotometer, and how just like UV-visible spectrophotometers, they split and compare light beams which pass through reference cells and sample cells to determine the transmittance of a sample.

We looked at molecular structures in a new light as well, not as a rigid construct of ideal bond lengths and angles, but as a dynamic, vibrating framework with many degrees of freedom for its atoms to move, and therefore, its bonds to stretch and bend. We covered how these vibrations tend to have frequencies corresponding to those of infrared photons, and these photons can be absorbed when they encounter a bond vibrating at that frequency, as long as the vibration causes a change in the overall dipole of the molecule.

We discussed factors affecting the rates of vibrations in both macroscopic systems and in bonds, how lighter atoms joined by tighter bonds lead to higher vibrational frequencies. Of course, this means that hydrogen-containing bonds, as well as double and triple bonds to larger atoms, tend to appear at the higher frequency end of the IR spectrum, called the functional group region, while vibrations in the carbon scaffold contribute more to the complex, but equally useful, fingerprint region in the lower frequencies.

Next time, we'll turn our attention to a phenomenon different from absorbance, a new form of interaction between light and matter called optical rotation. I'll see you then.

Measuring Handedness with Polarimetry
Lecture 29

I n this lecture, you will learn about the interaction between light and matter called optical rotation. You will learn that Étienne-Louis Malus is often credited for first observing plane-polarization of light from the Sun using a palace window and a piece of Icelandic spar as polarizers. Modern polarimeters operate on the same basic principles as in Malus's experiment, but the wavelength and plane of polarization of the analytical beam are more controlled. In addition, you will learn about the ability of chiral molecules to rotate plane-polarized light and how their specific rotation is equal and opposite for two enantiomers of the same compound.

Polarized Light

- Viking sailors as early as the first century B.C. claimed to have discovered a miraculous talisman known as "sunstone." Narratives about historical figures of that time suggest that this stone could be used to locate the position of the Sun even on the cloudiest of days. Whether or not such navigational techniques were used remains a bit of debate among archaeologists.

- There are those who believe that this apparently fictional and fantastic stone is actually a piece of Icelandic spar, also known as the mineral calcite. Calcite is unusual among common minerals because of a property called birefringence. Calcite has the ability to refract light at two different angles, depending on the orientation of the electromagnetic waves making up the incident light. Those waves oscillating perpendicular to the surface of the stone refract at an angle different than others, creating a second image when one looks through the stone.

- Scientists now know that light from the Sun that penetrates the Earth's atmosphere undergoes a process known as scattering, which happens when light encounters small particles in the air. Some familiar effects of this scattering are the blue color of the sky in the day and the reddish and orange hues of sunset. A second effect is polarization. The scattering that goes on as the light enters the atmosphere causes light that reaches the surface to have waves that are not randomly oriented, but rather polarized, or preferentially oriented in a direction that points back to the source—in our case, the Sun.

- Human eyes are not capable of detecting the difference between polarized light coming directly from the Sun and diffuse light coming from other directions, but with its strong birefringent properties, calcite is. By scanning the horizon looking though the stone and comparing the intensities of the two images one sees, a skilled navigator can pick out the direction of the Sun even on the most overcast of days.

- In 2002, a block of the mineral calcite was found in the wreckage of an Elizabethan ship called the *Alderney*, which sank in 1592. Perhaps most remarkable about this discovery is the fact that the stone was found just a few feet from other navigational tools.

Calcite has the ability to refract light at two different angles, making it able to differentiate between sunlight and diffuse light coming from other directions.

- Yet another tantalizing observation was made by physicist Guy Ropars at the University of Rennes in 2013. Ropars was able to demonstrate that, when properly trained in its use, a group of everyday people were able to determine the azimuth of the Sun on a cloudy day to within 1% of its true location.

- By the eve of the Industrial Revolution, some 300 years after the sinking of the *Alderney*, the technique seems to have been lost to the ages. We owe its rediscovery in the early 1800s to French soldier, engineer, and physicist Étienne-Louis Malus.

- In the wake of discoveries by Wilhelm Ritter and Sir William Herschel, Malus was contemplating the optical properties of minerals. One day, he was observing the optical properties of an Icelandic spar crystal outside the palace in Luxembourg. Malus noticed that if he looked at various objects through the crystal, he could see light of the same intensity regardless of how he rotated the crystal.

- But then he turned and faced the palace, and again rotated the crystal, looking at the light reflected by the palace's windows. What he saw was remarkable—that the intensity of the transmitted light changed as he rotated the crystal. Clearly, something about the process of reflecting sunlight was changing more than just the direction of its propagation.

- When non-polarized light strikes a smooth surface like a window at an angle that is specific to that material, some of that light is reflected from the surface, but some is also refracted through the surface. When the reflected beam and the refracted beam are at a 90° angle to one another, the reflected beam will be plane-polarized, with all of its constituent waves parallel to the orientation of that surface.

- What Malus saw through his Icelandic spar sample was simply his crystal either allowing that plane-polarized light through when its edge was properly aligned or blocking it from view when it was not.

- This is the principle on which polarized sunglasses work. Most polarized lenses are made by creating long, carefully oriented polymer molecules that are dyed to produce striations too small for the human eye to detect. These striations give the coating a polarizing effect. The coating is oriented on lenses so that it preferentially absorbs light that is plane-polarized parallel to the ground. In other words, it reduces the amount of road or ground glare you experience in your car or on the ski slopes.

The Polarimeter

- A modern polarimeter conducts an experiment not terribly different than the one conducted by Malus 200 years ago in France, although the parts are much smaller and the lamp life is less than 5 billion years, like our Sun. Malus used the Sun as a source of light, the windows of the palace as a polarizer, a crystal of Icelandic spar as a second polarizer, and his own eye as a detector.

- A modern polarimeter instead uses a source lamp—usually producing light in the ultraviolet or visible spectrum. An optical filter or monochromator ensures that only one wavelength at a time gets through to the polarizer. The polarizer, much like the windows of the palace, ensures that only one plane of polarization makes it to a sample compartment. So, at this point, we have plane-polarized light of a single wavelength encountering a sample.

- The path of the beam finally passes through a second polarizer on its way to a detector. If there is no change in the orientation of the plane of polarization, the rotational position of the second polarizer will have to be exactly the same as the first to maximize the amount of light reaching the detector. So, if we were to rotate the second polarizer, when this flash of light is observed, we know the plane of polarization of the light exiting the spectrometer.

Optical Rotation

- Small organic molecules in solution aren't aligned like polymer strands in your sunglasses. Instead, they tumble and move randomly, so they do not preferentially absorb plane-polarized light like a pair of sunglasses. They do, however, interact with plane-polarized light in a different way.

- Chiral molecules have the interesting ability to rotate plane-polarized light. As a plane-polarized light beam of a specific wavelength passes through a chiral sample, the plane of polarization rotates, so that when the beam exits the sample, the plane of polarization has a new orientation.

- The polarimeter must rotate the second polarizer to let the exiting light through to reach the detector. So, the instrument rotates the second polarizer until the detector sees the maximum intensity of light. The angle between the initial and final planes of polarization is called alpha. This is the measured rotation in a given experiment.

- This ability of chiral molecules to rotate plane-polarized light is called optical activity, and just as the ability to absorb light is an intrinsic property of a compound, so is its optical activity. In other words, a given chiral compound rotates a certain wavelength of light by a known and constant amount. This is called the specific rotation of a compound, and it is usually labeled as a bracketed alpha.

- Our observations so far lead us to derive an equation that is very similar to Beer's law. Just as absorbance is equal to the extinction coefficient multiplied by path length multiplied by concentration, rotation of an enantiopure sample is equal to its specific rotation multiplied by path length multiplied by concentration.

- But this is where polarimetry deviates from Beer's law. In UV-visible spectroscopy, there are no anti-absorbers that reemit the light absorbed by their counterparts, canceling them out. But in polarimetry, there is such a counterpart—the other enantiomer of the compound.

- We have to include a term in our equation that accounts for the opposing rotation of the minor enantiomer. Both terms of the equation share the same optical rotation and path length, so we can separate these algebraically and arrive at an equation that works for samples of a compound that are not enantiopure.

- So, observed rotation is equal to specific rotation multiplied by path length multiplied by the difference in concentration between (*R*) and (*S*) enantiomers. This difference is also called the enantiomeric excess. Journal articles covering chiral syntheses will always include a measure of the enantiomeric excess of the desired chiral product, reported as a percentage simply calculated by dividing the observed rotation by the calculated rotation and multiplying by 100%.

Polarimetry in Practice

- The technique of polarimetry had early and profound influence on the link between chirality and biological systems. Louis Pasteur famously used this technique in 1848 to analyze tartaric acid salt crystals isolated from wine. Pasteur's experiments showed very clearly that biologically sourced tartaric acid had the ability to rotate plane-polarized light but that tartaric acid synthesized in the laboratory did not.

- This lack of optical activity was a conundrum that lead Pasteur to take a very close look at synthetic tartaric acid. When he did, he noticed that carefully crystallized tartaric acid actually forms two different types of crystals. It was only when he painstakingly separated these crystals into two different groups by hand that he could recover the optical activity.

- This may be the very first example of a chiral separation. Fortunately, we have better methods today, but they all are dependent on Pasteur's realization that the same compound can have two equal but opposite optically active forms—enantiomers.

- But polarimetry has contributed even more to our understanding of organic reactions. One classical example of this is the work of Saul Winstein, who used polarimetry to recognize that not all S_N1 reactions proceed with total loss of stereochemistry, although most of them do.

- Winstein was able to show using polarimetry that the same S_N1 solvolysis reaction run in solvents of differing polarity not only proceeds at differing rates, but also with differing degrees of racemization. In other words, a chiral substrate might produce a racemic product when reacted in 50:50 water and alcohol but with 50% enantiomeric excess when run in alcohol alone.

- This led Winstein to theorize the existence of concerted ion pairs, in which positively charged carbocations and negatively charged leaving groups stay close to one another in what he called an intimate ion pair. If a nucleophile attacks before the ion pair fully dissociates, the leaving group (though detached) can still block the face of the carbocation from which it came. This explains the lack of total racemization in a reaction showing S_N1 kinetics, and the theory got its start with a polarimetry experiment.

Suggested Reading

McMurry, *Fundamentals of Organic Chemistry*, 6.3–6.4.

Wade, *Organic Chemistry*, 13.1–13.10.

Questions to Consider

1. Can the observed optical rotation of a sample ever exceed the calculated rotation? Explain.

2. Is it possible to synthesize enantiopure materials from achiral starting materials?

Measuring Handedness with Polarimetry
Lecture 29—Transcript

Viking sailors, as early as the 1st century B.C., may have discovered a miraculous talisman known as sunstone. Narratives about historical figures of that time suggest that this stone could be used to locate the position of the Sun even on the cloudiest of days. Whether or not such navigational techniques were used remains a bit of a debate among archaeologists. But there are those who believe that this apparently fictional and fantastic stone is not myth, but historical reality. They believe that it's this, a piece of Icelandic spar, also known as the mineral calcite.

Now, calcite is unusual among common minerals because of a property called birefringence. Calcite has the ability to refract light at two different angles, depending upon the orientation of the electromagnetic waves making up the incident light. Now, those waves oscillating perpendicular to the surface of the stone refract at an angle different than the others, and this creates a second image when one looks through the stone. This is a curious phenomenon, but what does it have to do a navigation, or for that matter, organic chemistry?

Scientists now know that light from the Sun, which penetrates the Earth's atmosphere, undergoes a process known as scattering. Scattering happens when light encounters small particles in the air. Some familiar effects of this scattering are the blue color of the sky in the daytime and the reddish and orange hues of sunset. But a second effect, one which most of us are probably not familiar with, is polarization.

The scattering that goes on as the light enters the atmosphere causes light, which reaches the surface, to have waves which are not randomly oriented, but rather are polarized, or preferentially oriented, in a direction which points back to the source, in this case, the Sun. Now, human eyes aren't capable of detecting the difference between polarized light coming directly from the Sun and diffused light coming from other directions. But with its strong birefringent properties, calcite is. By scanning the horizon, looking through the stone, and comparing the intensities of the images one sees, a skilled navigator can pick out the direction of the Sun on even the most overcast of days.

Now, the notion that the ancient Norse sailors used this technique to navigate is a very romantic tale. But the problem has been lack of direct evidence in the form of archaeological samples of calcite from sites dating to 2,000 years ago. There are those who believe that the calcite sunstones were used by these people and simply weathered away over the centuries, erasing all direct evidence that they ever existed. Others still, believe that the stories are pure fiction.

What is certain, however, is this, in 2002, a block of the mineral calcite was found in the wreckage of an Elizabethan ship, called the Alderney. Now the Alderney sank in 1592. Perhaps the most remarkable thing about the discovery is the fact that this stone was found just a few feet from other navigational tools, a tantalizing find. Yet another tantalizing observation was made by physicist Guy Ropars at the University of Rennes in 2013. Ropars was able to demonstrate that, when properly trained in it's use, a group of everyday people were able to determine the azimuth of the Sun, on a cloudy day, to within 1 percent of its true location.

If such a stone was ever used by the Vikings, or Elizabethan sailors, one thing is certain, by the eve of the Industrial Revolution, some 300 years after the sinking of the Alderney, the technique seems to have been lost to the ages. Now, we owe its rediscovery, in the early 1800s, to a French soldier, engineer, and physicist Etienne Louis Malus.

In the wake of discoveries by Ritter and Herschel, Malus, as the story goes, had taken an interest in optics and was contemplating the optical properties of minerals, like that Icelandic spar crystal I just showed you. One day, he was observing the optical properties of this beautiful material outside Luxembourg Palace. Now, what Malus noticed was that if he looked at various objects through the crystal, he could see light at the same intensity, regardless of how he rotated it. But then he turned and faced the palace, and again rotated the crystal, looking at the light reflected by the palace's windows. What he saw was remarkable, that the intensity of the transmitted light changed as he rotated the crystal. Clearly, something about the process of reflecting sunlight was changing more than just the direction of its propagation.

When non-polarized light strikes a smooth surface, like a window, at a very specific angle, and this angle is also, by the way, specific to the material, now, some of the light is reflected from the surface. But some of it is also refracted through the surface. When the reflected beam and the refracted beam are at a 90-degree angle to one another, the reflected beam will be plane polarized, with all of its constituent waves parallel to the orientation of the surface. And what Malus saw, through his Icelandic spar sample, was simply his crystal either allowing that plane-polarized light through, when it's edge was properly aligned, or blocking it from view when it was not.

So here is a sample of Icelandic spar, or calcite. And I have on the table in front of me a small image. Now, when I look at the image, without my Icelandic spar, I see, of course, a single copy. But placing the spar on top, I can see a very clear double image. The source of the double image is the fact that light of different planes of polarization is interplaying with the crystal in different ways, reflecting and refracting at different angles. Now, if there's any doubt about this, it can be easily proven using a modern polarizing filter, like the one I have here. If I look down through this filter at my Icelandic spar, at the image on the other side, simply rotating my plane-polarizing filter, I can make one image appear and the other disappear, and vice versa, simply by changing the arrangement of the plane of polarization, which is allowed through.

This is the principle on which polarized sunglasses work. Most polarized lenses, like the one I just used, are made by creating long, carefully oriented, polymer molecules, which are dyed to produce striations too small for the human eye to detect. These striations give the coating a polarizing effect. The coating is oriented on lenses so that it preferentially absorbs light, which is plane polarized parallel to the ground. In other words, it reduces the amount of road or ground glare you experience in your car or on the ski slopes.

So, we've seen how plane-polarized light can be generated by passing light through polarizing media, like the Earth's atmosphere, or, reflecting it from a flat surface. We've also seen how it can be detected by turning a second polarizer to either block, or allow, the plane-polarized light to pass through. But we're missing one last critical piece of the puzzle. How does plane-polarized light interact with organic samples? What critical questions can we answer with plane-polarized light that we couldn't answer otherwise?

Well, our tour of optical spectroscopy techniques has taken us through UV visible spectroscopy, which can be used to monitor energy difference and molecular orbital systems, and IR spectroscopy, which allows us to detect the presence of functional groups and also to fingerprint organic molecules using the distinctive signature vibrations of their carbon skeleton. So we have the tools to probe certain structural features of molecules with light, but we have not yet addressed the question of chirality. Of course, we know by now that chirality is a critical property of many organic molecules, from drugs like thalidomide to the proteins in your body. The profound importance of molecular handedness has been a focal point in many of our lectures.

We've discussed how Chris Ingold had a prolific run at characterizing the S_N1 and S_N2 mechanisms in the late 1920s, helping chemists to develop strategies to design syntheses which lead to products with the desired stereochemistry. But so far, we've not covered how Ingold could distinguish between the two different enantiomers and determine the relative populations in an otherwise pure sample.

The UV-visible and infrared spectra of RNS, thalidomide, for example, are expected to look exactly the same. Of course, determining whether a product is enantiopure, or racemic, or anywhere in between, for that matter, can be critical to our understanding of the mechanism which formed it. The key to determining the relative abundance of two enantiomers in an otherwise pure sample is a technique called polarimetry, which relies on one very interesting property of chiral molecules; they have the ability to rotate the plane of polarization in polarized light.

Now, a modern polarimeter conducts an experiment not terribly different than the one conducted by Malus 200 years ago in France, although the parts are much smaller and the lamp life is less than 10 billion years, like the Sun. Malus used the Sun as a source of light, the windows of the palace a polarizer, a crystal of Icelandic spar as the second polarizer, and his own eyes as the detector.

So the goal in our experiment is going to be to first create plane-polarized light and then to determine or verify the plane of its polarization. And to do so, all we need is two pieces of polarizing media and a light beam. Now, these are modern polarizing media; they are filters that can be used just like the coding of sunglasses, for example. So if I hold one of these up, you can still see through it pretty well. I'm still there, right? But the light that was going through this polarizing filter was exiting with only one plane of polarization, although, of course, the camera couldn't tell.

So in order to tell that I plane-polarized the light that's moving through my filter from my face, I can use a second polarizer. See, if I align them parallel to one another, the image makes it to the other side. But if, instead, I align them perpendicular, the light can't traverse the second polarizer. So, if I want to know the plane of polarization, I have to rotate my second polarizer until it is properly aligned with the first. This gives me a way to actually measure and detect the plane of polarization of polarized light.

Now, a modern polarimeter would use some type of a source lamp rather than the Sun. Frequently, we use sodium lamps, because they create just several different wavelengths of light themselves. Nonetheless, for our polarimetry experiment, we want only one wavelength of light to get into our experimental sample compartment. So, we'll place a monochromator after the lamp. In our case, we're going to use our monochromator to isolate just one wavelength of light, known as the sodium D line—589 nanometers.

All right, now, after this, I'm going to place a polarizer in line after the monochromator. The polarizer, much like the windows of the palace, ensures that only one plane of polarization makes it into my sample compartment. So, at this point we have plane-polarized light of a single wavelength encountering a sample. The path of the beam finally passes through a second polarizer on its way to a detector. So, if there's no change in the orientation of the plane of polarization, the rotational position of the second polarizer will have to be exactly the same as the first polarizer to maximize the amount of light reaching the detector. So, if we were to rotate the second polarizer, when the flash of light is observed by the detector, we know that the plane of polarization of light exiting the spectrometer can be measured.

Now small organic molecules in solution aren't aligned like polymer strands in sunglasses, though. Instead, they tumble and move randomly. So they do not preferentially absorb plane-polarized light like a pair of sunglasses. They do, however, interact with plane-polarized light in a different way. See, chiral molecules have the very interesting ability to rotate plane-polarized light. As a plane-polarized light beam of a specific wavelength passes through a chiral sample, the plane of polarization rotates so that when the beam exits the sample, the plane of polarization has a new orientation.

So let's go back to the polarimeter. So here we have my operating polarimeter, where my lamp, monochromator, and polarizer generate my monochromatic plane-polarized light, which traverse an empty compartment and exit through another polarizer, which is oriented in exactly the same angle as the first. So my detector's firing away, letting me know that the plane-polarized light is getting through.

Now, let's place a sample in our polarimeter. Now, placing a sample inside my polarimeter means that the beam will rotate the plane of polarization. Now, the beam itself is still going straight, but the plane of polarization of my beam is rotating. That means that my instrument has to rotate the second polarizer to let the exiting like get through and reach the detector. So my instrument has rotated that second polarizer until the detector sees a maximum intensity of light. We call the angle between the initial and final planes of polarization alpha. This is the measured optical rotation, and it's very simply measured by taking the first polarizer and the second polarizer and comparing the angle in between the two. We call this ability of chiral molecules to rotate plane-polarized light optical activity. And just as the ability to absorb light is an intrinsic property of a compound, so is its optical activity.

This is to say that a given chiral compound rotates a certain wavelength of light by a known and constant amount. We call this the specific rotation of a compound and usually label it as a bracketed alpha with 20 and D annotations, indicating that we're talking about the specific rotation of a compound at 20°C and in reference to that sodium-D emission line, which is 589 nanometers.

So, our observations, so far, lead us to derive an equation which is very similar to Beer's law. Just as absorbance is equal to the extinction coefficient, times the path length, times the concentration in Beer's law, so rotation of an enantiopure sample is equal to its specific rotation, times the path length, times concentration. So, if I have a solution of one gram of an enantiopure R chamfer dissolved in one milliliter of solvent in a one decimeter cell, knowing chamfer's specific rotation of 44.4 degrees per gram per mil decimeter, allows me to calculate that I should see a rotation of 44 degrees clockwise. Fantastic.

So let's say that I do just that with a pure sample of chamfer. I place the sample in the polarimeter. I collect my reading. And the rotation is only 22 degrees. So what just happened here? I have pure chamfer, but the observed rotation is less than the calculated rotation. What's going on here is that there's another form of chamfer, the S enantiomer, which rotates plane-polarized light with equal efficiency, but in the opposite direction because of its handedness.

So this is where polarimetry deviates from Beer's law. You see, in UV-visible spectroscopy, there are no anti absorbers which reemit light absorbed by their counterparts, thus canceling them out. But in polarimetry, there is such a counterpart. The other enantiomer of the compound. So for every degree that R-chamfer molecule rotates plane-polarized, D-line light, an S chamfer rotates it back to the opposite direction to the same extent, undoing that rotation. This explains why a pure sample of chamfer may rotate plane-polarized light less than expected. We have to include a term in our equation which accounts for the opposing rotation of the minor enantiomer. Now, both terms of the equation share the same optical rotation and path length. So we can separate these algebraically and arrive at an equation which works for samples of a compound which aren't enantiopure.

Here we have observed rotation equal to the specific rotation of the compound, times the path length, times the difference in concentrations between R and S enantiomers. This difference is also called the enantiomeric excess. Journal articles covering chiral syntheses will always include a measure of the enantiomeric excess of the desired chiral product, reported as a percentage simply calculated by dividing the observed rotation by the calculated rotation and multiplying by 100%.

Now, it's worth pausing here to note that the enantiomeric excess of one stereo isomer is not the same as its total abundance in the sample. In our sample we had an observed rotation of 22 degrees and a calculated rotation of 44 degrees. This makes the enantiomeric excess, or EE, of the R chamfer in our sample 50%. But the total amount of R chamfer in our sample is that excess, plus half of the remainder, since the remaining 50% of our sample were molecules of chamfer, which were canceling one another out. So the total abundance of the R enantiomer is 50%, plus one half of the remaining 50%, for a total of 75% R enantiomer. Now, in general, when organic chemists report their results in an article, it's assumed that the reader can conduct this manipulation on their own, so we usually see just a report of EE and not one of total abundance.

The technique a polarimetry had early and profound influence on the link between chirality and biological systems. Louis Pasteur famously used this technique in 1848 to analyze tartaric acid salt crystals isolated from wine. Pasteur's experiments showed very clearly that biologically sourced tartaric acid had the ability to rotate plane-polarized light. But tartaric gas it synthesized in the laboratory did not. So, in fact, what was going on here was the term racemic was coined this way because it refers to an equal mixture of enantiomers, and it's derived from the Latin for bunch of grapes, because of this observation, a testament to the role of tartaric acid in our understanding of chirality.

This lack of optical activity was a conundrum which led Pasteur to take a very, very close look at synthetic tartaric acid. When he did, he noticed that carefully crystallized tartaric acid actually forms two different shapes of crystals. It was only when he painstakingly separated these crystals into two different groups, by hand, that he could recover the optical activity of tartaric acid. Crystals of one shape could be combined to produce the optical activity of the tar traits recovered from wine production. And the other, quite interestingly, though not surprising to us now, show equal and opposite activity.

So this may be the very first example of a chiral separation. Fortunately, we have better methods today than tweezers and careful observation that Pasteur had to use. But all of these methods find their origins in Pasteur's realization that the same compound can have two equal, but opposite, optically active forms or enantiomers.

But polarimetry has contributed even more to our understanding of organic reactions. One classic example of this is the work of Saul Winstein, a Canadian chemist with a deep interest in carbocations. Winstein spent much of his career investigating situations in which this familiar reaction intermediate seemed to act in ways that it shouldn't. One of these situations takes place in certain S_N1 reactions. Winstein used polarimetry to recognize that not all S_N1 reactions proceeded with a total loss of stereochemistry, even though most of them, in fact, do. Winstein was able to show, using polarimetry, that the same S_N1 solvolysis reaction run on a chiral substrate in solvents of different polarity, not only proceeds at differing rates, but with differing degrees of racemization. This is to say that a chiral substrate might produce a racemic product when reacted in higher polarity solvents, but with maybe a 50% EE when run in lower polarity environments, in which a carbocation would be less stable.

This led Winstein to theorize the existence of concerted ion pairs in certain reactions, in which positively charged carbocations and negatively charged leaving groups still form but stay close to one another in, what he called, an intimate ion pair. So, if a nucleophile attacks before the ion pair fully separates in space, the leaving group, though detached, can still block the face of the carbocation from which it came. Now, this explains the lack of total racemization in a reaction showing S_N1 kinetics. And it changed our way of thinking about the S_N1 and S_N2 mechanisms.

See, thanks to Winstein's work and the technique of polarimetry, we realize now that nucleophilic substitution mechanisms are not necessarily an idealized S_N1 or S_N2, but rather, a continuum of mechanisms ranging from the classical S_N1 with total racemization through a S_N1 involving ion pairs and having partial conversion of chirality, to classical S_N2 with complete retention of chirality.

So let's review what we've covered today. We started with a bit of history, discussing how Etienne Louis Malus is often credited for first observing plane polarization of light from the Sun using a palace window and a piece of Icelandic spar as polarizers, although Viking navigators may have actually beaten him to it by about 2,000 years. We saw how modern polarimeters operate on the same basic principles as in Malus's experiment, but the wavelength and plane of polarization of the analytical beam are more controlled.

We investigated the ability of chiral molecules to rotate plane-polarized light, and we saw how their specific rotation is equal and opposite for two enantiomers of the same compound. It's this equal-and-opposite relationship which allows us to observe the optical rotation in an otherwise pure sample, comparing it to the calculated optical rotation of the sample to determine the enantiomeric excess, remembering that the enantiomeric excess is not the same as total abundance.

Finally, we took a look at the work of Louis Pasteur and Saul Winstein, two scientists separated by century, but both of whom were able to draw critical inferences about chirality in organic compounds by observing the interaction of plane-polarized light with those compounds.

Next time, we will move to yet another region of the electromagnetic spectrum. We're going to take a look at the phenomenon of nuclear magnetic resonance and how tiny atom-sized magnets in organic molecules interact with radio waves and one another to produce an effect which has become one of the gold standard identification tools of modern organic chemistry. I'll see you then.

Nuclear Magnetic Resonance
Lecture 30

In this lecture, you are going to learn about yet another region of the electromagnetic spectrum. You will go all the way to the long-wavelength end of the spectrum—beyond infrared and past microwaves all the way to the radio portion of the spectrum. But before doing so, you need to better understand a property of atoms that is usually thought of as more within the realm of physics than of organic chemistry: magnetism.

Magnetism

- The magnetic properties of nuclei were not well understood by Sir Isaac Newton, Ritter, and Sir William Herschel and went largely unaddressed during the development of optical spectroscopy techniques in the 1800s. But in the early 1900s this changed, in large part due to the efforts of Isidor Rabi, an American physicist who is credited with having the revelation that atoms have certain magnetic properties that can be used to discern one from another in a given molecule.

- Atomic nuclei have a property that physicists call magnetic "spin"—a term that does not actually refer to the rate of rotation, as one might assume from classical mechanics. Certain atomic nuclei have a spin quantum number of 1/2, which is just a fancy way of saying that there are two possible magnetic states. Such nuclei contain what is known as a magnetic dipole, or a magnetization that can behave much like an atom-sized bar magnet or compass needle.

- Isotopes, or atoms of the same element with different numbers of neutrons, have different magnetic properties. Nearly every element has at least one isotope that is spin 1/2, including protons, carbon 13, nitrogen 15, and oxygen 17.

- The magnetic dipole of certain atomic nuclei behaves in much the same way as a compass needle—randomly orienting itself in the absence of an external magnetic field, but as soon as a magnetic field is applied, specific orientations of the nuclear dipole become more stable.

- Unlike a compass needle, however, spin 1/2 atomic nuclei can align their dipoles either parallel to or antiparallel to a magnetic field in which they reside, leading to a distribution of these two states in the population.

- But the energies of the parallel state (called the alpha spin state) and the antiparallel alignment (called the beta spin state) differ by an amount that can be calculated and is referred to as the Zeeman splitting energy, which is a function of the intensity of the external magnetic field and an intrinsic property of the nucleus called the gyromagnetic ratio. All protons in the universe have the exact same gyromagnetic ratio.

- So, for a nucleus of a given type, the stronger the external magnetic field, the greater the energy difference between alpha and beta spin states will be.

Continuous Wave Nuclear Magnetic Resonance

- Early spectrometers consisted of a powerful electromagnet, a radio frequency transmitter, and a radio frequency receiver. A sample of organic material is placed into the magnetic field and irradiated with a constant wave of a specific radio frequency. This is how "continuous wave" nuclear magnetic resonance (NMR) gets its name.

- The current to the electromagnet is slowly increased as the exiting radio wave is monitored. At one very specific field strength, the radio wave will be absorbed in the spin state transition from alpha to beta, manifesting itself as a reduced radio wave intensity at the receiver. This is the basic design of a continuous wave NMR spectrometer.

Chemical Shift

- If all protons have the exact same gyromagnetic ratio and the field in an NMR experiment is uniform, then how is it that this technology is good for anything other than simply detecting the presence of protons? How can it possibly distinguish one proton from another? The answer to this very important question lies in a different term in the Zeeman splitting equation. It is not the gyromagnetic ratio that differs from one proton to the next but, rather, the exact strength of the external field at each nucleus.

- The key to understanding how NMR gives us information lies in another phenomenon—magnetic shielding by electrons. Electron clouds have this interesting ability to attenuate, or weaken, the external field ever so slightly. We can think of this as a new term in the Zeeman equation in which we account for a small reduction in field strength due to shielding.

- A proton in an alkyl group will have a certain amount of electron density around it, ever so slightly reducing the field strength at its nucleus, whereas a proton closer to an electron-withdrawing atom is expected to have less electron density around itself.

- Reduced shielding means that we need to apply a less-powerful field to bring a given proton into resonance with the applied radio-frequency beam.

- The density of the electron cloud at a particular nucleus can cause its peak to shift within the NMR spectrum. This phenomenon is called chemical shift.

- Because they resonate at higher applied fields, more-shielded protons appearing to the right of the spectrum are called upfield. The converse—protons with little shielding—are said to resonate downfield.

- A final factor that must be considered when predicting chemical shifts is the phenomenon of magnetic anisotropy, which is the tendency of electrons in pi systems to be held less tightly and move through a greater volume of space than their sigma counterparts. This leads to a situation in which the electrons behave almost like a small solenoid, acting like a current moving through a wire, generating a magnetic field of their own.

The Standard Unit of NMR Spectroscopy

- Protons in different chemical environments will resonate at distinct applied fields because of effects like shielding. But what happens when a researcher in a lab across the country or across the globe tries to reproduce your experiment? What happens if the researcher's spectrometer produces a field strength or radio wave frequency that is different than yours?

- All shifts of all of the protons will be different, making comparison somewhat difficult. So, we need a method of reporting data that can be easily translated from one spectrometer to the next. We need to normalize the resonance frequencies to account for different applied field strengths.

- We start this process by agreeing to a reference standard. In the case of proton NMR, we use a compound called tetramethylsilane (TMS). We use this compound because the very low electronegativity of silicon pushes electrons toward the methyl groups, producing a strong, consistent resonance frequency that is so far upfield that it rarely interferes with the signals of interest.

- The difference in field needed to produce resonance in the protons of most organic molecules is just a small fraction of 1% of the total field strength. To make the numbers more manageable, we report chemical shift in parts per million from the absolute resonance frequency of that reference compound TMS. This also allows us to compare data collected on one instrument with data collected on another instrument using a different magnetic field strength. So, parts per million (ppm) has become the standard unit of NMR spectroscopy.

Magnetic Coupling

- How would we expect a continuous wave NMR spectrum to look for a compound that has more than one kind of proton? Let's consider this using a fairly simple organic compound, ethyl chloride, which has two chemically distinct types of protons: those on the carbon bearing the electron-withdrawing chlorine and those one carbon-carbon bond away.

- We expect the former to be more deshielded than the latter, producing a spectrum with two different absorptions: one at about 3.5 ppm and one at about 1.5 ppm. This is, in fact, what we see. But if we look closer, we see that the peak at 3.5 ppm appears to actually have four distinct lobes and the peak at 1.5 has three.

- This is a phenomenon known as spin-spin coupling, a mechanism through which magnetization from one neighboring proton is encoded on the other. This phenomenon is most obvious when the coupled hydrogens are three bonds away from one another and only takes place when the chemical shifts of the coupled protons are different.

- NMR spectroscopists are familiar with the patterns produced by the interplay of nuclei close to one another in a molecule and look for distinctive patters that help them identify not only the presence, but also the location of groups within a molecule.

Exchangeable Protons

- More than any other element, hydrogen atoms (or protons) tend to exchange between molecules. For example, the acid proton of a carboxylic acid is very labile, potentially protonating molecules of solvent or even other carboxylic acid molecules in the sample. This transition between and among different chemical states can occur on the timescale of an NMR experiment, which usually takes a few seconds to complete, leading to a "blurring" of the chemical shift associated with that proton.

- A collection of molecules with protons interchanging very slowly will produce spectra with sharp lines for each type of proton, but as their exchange rate increases, the signals coalesce and blur. This explains why protons bonded to nitrogen and oxygen have a tendency to appear at many different chemical shifts and with broader absorption peaks than usual.

- This sort of behavior is very common for acid, alcohol, and amine protons, and it explains why their approximate chemical shifts are so widely varied in shift tables.

Suggested Reading

McMurry, *Fundamentals of Organic Chemistry*, 13.7–13.14.

Wade, *Organic Chemistry*, 12.13–12.15.

Questions to Consider

1. Carbon-13 has a gyromagnetic ratio about 1/4 that of a proton. Will carbon-13 nuclei resonate upfield or downfield of protons?

2. Why is it critical for the magnetic field within an NMR spectrometer to be perfectly uniform across the entire sample during the experiment?

3. Most proton NMR experiments are run using solvents containing deuterium instead of hydrogen. Why?

Nuclear Magnetic Resonance
Lecture 30—Transcript

We've spent some time now learning about the discovery of how electromagnetic waves in and near the visible spectrum can interact with organic molecules and how we can observe that interaction in a way which gives us information about the structure of the molecules involved.

We learned how Newton's use of a prism to separate light into its constituent wavelengths introduced the Western world to the idea that visible light was made up of multiple color components. We saw how the discoveries of Herschel and Ritter demonstrated that there were other types of light just outside of that visible spectrum as well. We've talked about how all three of these portions of the spectrum—ultraviolet, visible, and infrared—interact with small molecules, and the tremendous utility that these phenomena have as tools for the identification of organic compounds.

So far, we've investigated how to determine the extent of conjugation in a compound with UV-visible spectroscopy, the presence of functional groups using IR spectroscopy, and the handedness of compounds using polar imagery. But short of comparing the fingerprint region of IR spectra to those of known standards, we don't really have a way to probe the connectivity functional groups within a molecule.

So, if we're making an entirely new molecule for which there is no reference spectrum yet in the literature, how are we going to confirm its identity? In this lecture, we will move on to yet another region of the electromagnetic spectrum. To get our answer, we're going all the way to the long-wavelength end of the spectrum—beyond infrared, past microwaves, all the way to the radio portion of the spectrum.

But before we get to our radio waves today, we'll need to try to better understand a property of atoms that we haven't really discussed and which we usually think of as more within the realm of physics than organic chemistry. That property is magnetism. Now, the magnetic properties of nuclei were not well understood by Newton, Ritter, and Herschel, and they went largely unaddressed during the development of optical spectroscopy techniques in the 1800s. But in the early 1900s, that all changed. In large part, due to the efforts of Isidore Rabi, an American physicist who is credited with having the revelation that atoms have certain magnetic properties which can be used to discern them from one another in a given molecule. So, to get started today, let's take one more step back and consider magnetism in general.

But, before we get to our radio frequency radiation, I'd like to show you this. It's a compass. It's a device with which most of us, especially navigators, pilots, hikers, are familiar. A small magnetized needle is mounted on an axis which allows it to turn so that it can align itself with whatever sort of magnetic field it's in. Of course, in the case of my compass needle, this field is generated by the Earth's rotating inner core.

Now, physicists have several different units used to measure the intensity of a magnetic field, one of which is the Tesla, named for the famous and somewhat eccentric physicist Nikola Tesla. At it's surface, the Earth's magnetic field measures about 50 milli Tesla. And here close to Washington D.C., this weak but ever present field points about 11 degrees to the west of true north. So when I place this compass within the field, the needle aligns itself with that field, giving me an indication of its direction.

So let's assume for a moment that I could somehow remove the influence of the Earth's magnetic field. Now, I can't remove the Earth's magnetic field, because I can't stop the core from spinning. But what I can do is use these rare Earth magnets to pull this needle out of alignment. Now, if there were no magnetic field, I would expect that any arrangement whatsoever of this needle in space would be of equal energy. So any alignment is possible. But as soon as I remove my rare magnet and allow the Earth's magnetic field to take over once again, this guy back out of alignment, snaps right back. So, as soon as I place my compass back in the field, the energy of the aligned state becomes lower in energy than any of the other misaligned states, causing the needle to reliably and predictably take on that lower energy orientation.

Now, the reason I'm showing you this, this rather mundane little magnetic tool, is that it's a very good analogy for the behavior of certain atomic nuclei. Atomic nuclei have a property that physicists call magnetic spin. I specifically mention physicists here because I refuse to lay the blame for this one on chemists. Spin, you see, is a very unfortunate choice of terminology, as it doesn't actually refer to a rate of rotation as one might assume from classical mechanics. It's a common point of confusion in the study of NMR, so let's be sure we clear that up right away.

Now, for our discussion, it'll suffice to say this, certain atomic nuclei have a spin quantum number of one half, which is just a fancy way of saying that there are two possible magnetic states. Such nuclei contain what is known as a magnetic dipole, or a magnetization which can behave much like an atom-sized bar magnet or compass needle.

Finally, those neutrons that I encouraged you to overlook way back at the beginning of the course are about to become relevant. This is because isotopes, or atoms of the same element with different numbers of neutrons, have different magnetic properties. And nearly every element has at least one isotope, which is spin one half. This includes protons, carbon 13, nitrogen 15, and oxygen 17. But for this lecture, we're going to confine ourselves to protons with the caveat that the same magnetic tricks I'm about to show you can be pulled with at least one isotope of practically every commonly encountered element.

So, let's take a look at a hydrogen nucleus. The magnetic dipole of certain atomic nuclei behaves in much the same way as our compass needle did, randomly orienting itself in the absence of an external magnetic field, but as soon as a magnetic field is applied, which we represent as these lines, specific orientations of the nuclear dipole become more stable.

Now, where the analogy deviates from our macroscopic example of our compass and needle is that spin one half atomic nuclei can align their dipoles either parallel to or antiparallel to a magnetic field in which they reside, leading to a distribution of those two states in the population.

But the energies of the parallel state, often called the alpha spin state, and the anti parallel alignment, sometimes called the beta spin state, differ by an amount which can be calculated, and is referred to as the Zeeman splitting energy. The Zeeman splitting energy as a function of the intensity of the external magnetic field and an intrinsic property of the nucleus called it's gyromagnetic ratio. Now, by intrinsic I mean that, for example, all protons in the universe have the exact same gyromagnetic ratio. So, for a nucleus of a given type, the stronger the external magnetic field, the greater the energy difference between the alpha and beta spin states will be.

Now, on our way to understanding how NMR works in laboratories of the 21st century, we first have to make a pit stop in the 1940s and '50s, when the science was still being developed by Rabi and his contemporaries. This historical context is going to help us to better understand the more complex instruments of today.

So let's say that we have a collection of protons in a fairly strong external magnetic field. We expect that these protons will arrange themselves in a collection of alpha and beta spin states with just a very slight excess in the lower energy alpha state. Let's say that we place our protons in a 1.4 Tesla magnetic field. That's a field about 28,000 times as strong as the Earth's, powerful, but attainable with a suitable electromagnet. Based on the gyromagnetic ratio of a proton, 42.577 megahertz per Tesla, I can calculate the Zeeman splitting energy difference between those two states. Since the proton is the most frequently used nucleus in NMR experiments, we report field strengths in Tesla, or rarely in Tesla, more often, we actually use the resonance frequency of a proton in that field.

For example, our 1.4 Tesla magnet should bring a proton into resonance with a radiation of 60 megahertz in frequency. So, you'll notice that most modern spectrometers have a frequency splashed across the side in large fonts. This is the frequency that's indicated there. It's a photon of that specific frequency, one with a wave length of about five meters, which carries just the right amount of energy to cause a proton in the alpha spin state to flip into the beta spin state, absorbing the photon in the process.

So, early spectrometers attempting to take advantage of this phenomenon consisted of a powerful electromagnet, a radio frequency transmitter, and a radio frequency receiver. A sample of organic material is placed into the magnetic field and irradiated with a constant wave of a specific radio frequency. In keeping with our example, that would be 60 megahertz. And this is how continuous-wave NMR gets its name.

The current to the electromagnet is slowly increased as the exiting radio wave is monitored at the receiver. In our very simple example, at one very specific field strength, the radio wave will be absorbed in the spin-state transition, from alpha to beta, manifesting itself as a reduced radio wave intensity at the receiver. So this is the basic design of a continuous-wave NMR spectrometer.

So, if you've been following so far, you probably have at least one very important question. If all protons have exactly the same gyromagnetic ratio and the field in the NMR experiment is perfectly uniform, how is it that this technology is good for anything other than simply detecting the presence of protons? How can it possibly distinguish one proton on a molecule from another? The answer to this very important question lies in a different term in the Zeeman splitting equation. It's not the gyromagnetic ratio which differs from one proton to the next, but rather, the exact strength of the external field at a given nucleus.

You see, we started with the very unrealistic example of a collection of isolated protons with no other atoms or even electrons present in the system. The obvious difference between this model system and the ones which a chemist might want to study is that there are electron clouds and atoms of other kinds bonded to the protons that we want to analyze. And the real key to understanding how NMR gives us information lies in another phenomenon—magnetic shielding by electrons.

You see, I told a little white lie back there, at least a lie of omission. I already told you the Zeeman splitting is a function of gyromagnetic ratio and the external field strength. But what I left out was the fact that electron clouds have this interesting ability to attenuate or weaken the external field ever so slightly. Now, we can think of this as a new term in the Zeeman equation in which I account for a small reduction in field strength due to the shielding of the electron cloud at that particular nucleus.

Let's take a look at some examples to see if we can drive this point home. We're going to look at some hypothetical NMR spectra for four different compounds—methane, chloromethane, dichloromethane, and chloroform. So, essentially, what I've done here is started from methane and then slowly added one chlorine atom at a time, depicted here in green, until I reach the point where I have three chlorines attached.

So, let's start by thinking about methane and its electron cloud. Now, the electron cloud in a methane molecule is very symmetrical, isn't it? There are no particularly electronegative elements. The carbon hydrogen bonds are essentially nonpolar. And so, there really is no distortion in this electron cloud. That means that if I plot its absorption versus the applied field strength in a continuous-wave NMR spectrometer, I expect that it will take a fairly high field to overcome the shielding that these protons are experiencing.

But if I move along to, say, chloromethane, where I've replaced one of those hydrogens with a more electronegative chlorine atom, there's now a dipole within this molecule and it's pulling electrons toward the chlorine, and therefore, away from the remaining hydrogens. So its electron cloud may look more like this with a bit of a distortion, because that chlorine is pulling some of it more towards itself. So, with reduced shielding, I expect that I need a lower applied field strength to create resonance. And indeed, that's what I see. And this trend continues to dichloromethane, which has two dipoles, and therefore, an electron cloud which is even more deshielded at those protons. And of course, its NMR spectrum would reflect that in an even weaker applied field necessary for resonance.

And finally, chloroform, where I have three of these dipoles, has got a very, very deshielded proton, which will resonate at an even lower applied field. Because of this, lower applied fields causing resonances down here at the left-hand side of the spectrum, we refer to this side of the spectrum as downfield. Whereas we refer to the higher field portion of the spectrum sometimes as upfield. So a more shielded hydrogen, like those in methane, resonates upfield; a more deshielded hydrogen, like those in chloroform, will resonate downfield.

So the density of the electron cloud at a particular nucleus can cause it's peak to shift within the NMR spectrum. So we call this phenomenon chemical shift. Now, because they resonate at higher applied fields, more shielded protons appearing to the right of the spectrum are called upfield. And the converse, those with little shielding, are said to resonate downfield. So in our example, a spectroscopist might say the methyl protons resonate upfield of the methyl chloride protons.

So we've established that protons in different chemical environments will resonate at distinct applied fields because of the effects like shielding. But what happens when a researcher in a lab across the country or across the globe tries to reproduce my experiment? What happens if his or her spectrometer produces a field strength or radio wave frequency different than mine? Well, of course, all shifts of all the protons will be different, making comparisons somewhat difficult. So we need a method of reporting data which can be easily translated from one spectrometer to the next. We need to normalize the resonance frequencies to account for different applied field strengths. And we start this process by agreeing to a reference standard.

In the case of proton NMR, we use a compound called tetramethylsilane. We use this compound because the very low electronegativity of silicon pushes electrons toward the methyl groups, producing a strong, consistent resonance frequency which is so far upfield that it rarely interferes with the signals of interest. So the differences in field needed to produce resonance in the protons of most organic molecules is just a small fraction of 1% of the total field strength.

So to make the numbers more manageable, we report chemical shift in parts per million from the absolute resonance frequency of that reference compound, TMS, or tetramethylsilane. So a proton resonating in a field which is 99.9999% of the strength needed to bring TMS into resonance is said to be resonating at 1.0 PPM chemical shift. A proton at 99.9998% of the field strength of TMS would be at 2.0 PPM chemical shift, and so on.

Now, treating the x-axis of my spectrum in this way not only gives me easier numbers to work with, but allows me to compare data collected on one instrument with data collected on another instrument, even if it uses a different magnetic field strength. So parts per million has become the standard unit for NMR spectroscopy.

So we've considered the behavior of a hypothetical collection of protons in a magnetic field. We then graduated to real, albeit very simple, organic compounds, like methane and its halogenated derivatives, all of which have only one type of proton in them. But what happens when a compound has more than one kind of proton? How would we expect a continuous-wave NMR spectrum to look for a compound like that?

Let's answer that question with an example. Let's take a look at ethylchloride, or chloroethane. Now, chloroethane has two chemically distinct protons. These I've highlighted in red are adjacent to the chlorine atom. And of course, there are three others are farther away. Now, if I locate the resonance frequency for these two protons, I notice something very interesting has taken place. It's not a single peak, even though these are identical and should produce a single absorption. I actually see a peak which has four distinct lobes. Now, the origin of these lobes is the interaction of these two protons with their neighbors. See, every proton, remember, is a small bar magnet, and that means that whether they're aligned parallel to or anti-parallel to the applied magnetic field will cause them to interact differently with the adjacent protons.

So if I consider the effect of these protons which I've highlighted in blue, they can interact with the hydrogens of interest by being all spin up. There are actually three different ways that I can combine them with two spins up and one spin down. And of course, there will also be three potential ways to combine them with two spins down and one spin up. And the final permutation that's possible with three is to have all spin down.

So in the case where all of the neighbors are spin up, they add just a little bit of a magnetic field to what's experienced by those causing the signal, and that means it requires just a little bit less of an applied magnetic field to cause resonance. That's why what should be, in theory, a single signal, is actually split into four. So these are the source of that multiplicity.

Similarly, I can look at the resonance frequency of these three hydrogens and say they must be coupled to or interacting with my red hydrogens. Now, with two hydrogens interacting, I have the possibility of having them both spin up, two possibilities in which I have one up one down, and my final possibility in which they're both spin down.

So, in this case, my blue hydrogens should create a resonance with three lobes. So we refer to this phenomenon as magnetic coupling, and the resulting shape of the peaks as multiplets. And they follow a rule known as the n plus 1 Rule, which simply states that we have n plus 1 lobes per multiplet, where n is the number of coupled hydrogens. So, my red hydrogens are coupled to three neighbors, creating a peak with four lobes. My blue hydrogens are coupled to two neighbors, creating a peak with three lobes. In most simple situations where we're dealing with saturated compounds with free rotation, this rule holds pretty well.

So let's take a look at some multiplet features that we may see in NMR spectra as we go forward. Of course, the simplest where n equals 1 we call a doublet. When there are two coupled partners, we get what we call a triplet. Three produces a quartet, and so on, with four making a quintet, and five making a sextet, and so on. Now, in many cases they become so complex around this point that we stop referring to them by their individual names and simply refer to them collectively as multiplets.

Now, the intensity of the lobes within the multiplet is also useful information, so we'd like to be able to predict that as well. And we can. For example, a doublet is always one to one; whereas a triplet is one, to two, to one; and a quartet one, to three, to three, to one. And the pattern continues. Remember, this is because of all the different permutations of alpha and beta spin states for the coupling partners.

Now, what's really great is if you forget this pattern, you can easily recreate it using a geometric mnemonic. That geometric mnemonic is known as Pascal's Triangle. So we can simply create a series in which we add adjacent numbers as we go down, so one plus two equals three. For example here, three plus three equals six, etc. And Pascal's Triangle allows us to predict the multiplicity in any instance where we have multiple coupling partners in simple situations like these.

So we're now ready to get just a small taste of the power NMR has to reveal structural information. We're going to begin with a very simple example, but one which clearly shows how we can gather structural information which would not be available to us from the UV vis, IR, or polar imagery experiments that we have covered so far. And we're going to do it using a relatively simple molecule, or set of molecules, and that is the structural isomers of chloropropane.

So let's construct some hypothetical NMR spectra for the two different isomers of chloropropane. We'll start with two chloropropane. So here I have my chlorine attached at the central atom. And because of this, I'm going to have a spectrum which has two different resonances. Now, it's worth mentioning here I've exaggerated the multiplicities a little bit. I've spaced them out a little bit wider than you might see them in an actual spectrum, so that they're easy for us to see. But their position within the spectrum is accurate.

So we see that there are two different kinds of protons in this two chloropropane. First, there are the six methyl protons, which are all exactly the same as one another. Now, because they're chemically identical to one another, they don't couple to each other, but they do have one coupling partner in this proton right here in the center of the molecule. And it has its own resonance, which is split by its six non-equivalent neighbors.

So, just as we would expect from our previous discussions on multiplicity, we have a very large doublet, which corresponds to these six hydrogens highlighted in blue, and a relatively small multiplet, produced by this single hydrogen, which is split by many coupling partners. Now, if we take a closer look, we can see the multiplicity is exactly as we would predict using Pascal's Triangle, and that this is actually a one, two, three, four, five, six n plus 1. This is a septet. So our septet at 3.7 parts per million tells us we have a proton with many coupling partners very close to an electron-withdrawing group. And our doublet it at 1.1 parts per million tells us that we have some protons which have a single coupling partner, and that those protons are farther away from the electron-withdrawing group because they resonate upfield.

So how can we use this spectrum in comparison to, say, one chloropropane and distinguish between these two compounds? Let's take a look at one chloropropane right now. So when I put my chlorine at the end of the molecule, rather than in the center, it has a very important and profound effect on the number of resonances that I see. My spectrum is a little bit more complicated now, isn't it? That's because by placing the chlorine at the end, I've created three distinct chemical environments for the protons in the molecule. Those that are directly on the carbon bearing the chlorine, those one bond away, and those that are two bonds away. So I expect to see three distinct resonances, which I do. But not only that; their multiplicities are different than those in my previous example.

So, in this case, I have a triplet that is way upfield, which tells me I've got some hydrogens that are very far away from any electron-withdrawing groups, and, have two non-equivalent neighbors to which they are magnetically coupled. I also have a little bit more complicated multiplet here. This one would be called a sextet, and that tells me that I've got some protons which are a little bit closer to the electron-withdrawing group and are coupled to five distinct partners—one, two, three, four, five.

Finally, I have another triplet which is farther downfield, telling me that I must have some protons that are very close to the electron withdrawing chlorine and are coupled to two partners, which is a perfect description of these two protons highlighted in red, coupled to their neighbors in blue. So this spectrum having a triplet at 0.8, a sextet at 1.6, and another triplet at 3.3 allows me to very clearly and unambiguously assign that it's one chloropropane.

So, let's review the material from this lecture. We started by considering a property of atomic nuclei known as spin and how certain isotopes of certain atoms have a spin of one half, giving them two possible spin states when an external magnetic field is applied. There's the low energy parallel state called alpha, and the high energy anti-parallel state called beta.

We discussed how the energy difference between these states, known as the Zeeman splitting, corresponds to the radio frequency region of the electromagnetic spectrum and is proportional to the strength of the magnetic field with which the nuclei are aligning themselves.

We took a look at the design of early spectrometers, in which a continuous radio frequency wave is shined on a sample as the strength of an applied field is varied until resonance is achieved, resulting in an absorption of the radio wave.

We then looked at how the shielding and induced fields provided by electrons can cause the magnetic field at protons within a molecule to be slightly different, leading us to the calculations of chemical shift.

Finally, we discussed the phenomenon of magnetic spin-spin coupling, in which non-equivalent nuclei, which are three bonds away from one another, can share magnetization, splitting residences into a set of lobes called a multiplet.

Next time, we'll take our discussion into the new millennium and look at what is known as pulsed, or Fourier transform NMR. We'll also take a brief look at a few other techniques used to characterize molecules, specifically X-ray crystallography and mass spectrometry. I'll see you for that discussion next time.

Advanced Spectroscopic Techniques
Lecture 31

———————————————————————————

This lecture will take our discussion into the new millennium by analyzing what is known as pulsed, or Fourier-transform, NMR. You will also learn about a few other techniques used to characterize organic molecules—specifically, you will learn about X-ray crystallography, which has been used to validate many of the structural theories that preceded it, and mass spectrometry. In addition, you will discover that we must always remember that these techniques have their limitations.

Pulsed NMR

- Several advances in technology have led to the development of the pulsed, or Fourier-transform, NMR technique. First and foremost is the development of superconductors. All modern high-field spectrometers use a superconducting magnet to generate extremely high applied fields to achieve better resolution.

- Keeping these superconducting magnets cold is critical to their function, so spectrometers contain a series of concentric vacuum-walled compartments called dewars—an exterior dewar filled with liquid nitrogen and an interior dewar containing liquid helium and the superconducting magnet. The liquid helium keeps the magnet at 4 kelvins—colder than the dark side of the Moon—while just inches away from it is the sample at room temperature.

- The second advancement that opened the door to the pulsed NMR method is computing technology. The detector in NMR is more often called a receiver because the response that it creates is not a spectrum but, rather, a raw data set that must be manipulated by the computer before it can be used.

- Recall that in the presence of a magnetic field, protons align their dipoles in the parallel alpha spin state or the antiparallel beta spin state, and the populations of these two states are governed by the

Zeeman splitting, which is the energy difference between the alpha and beta states and can be calculated using the field strength at the nucleus and the gyromagnetic ratio.

- There are two additional principles that help explain the pulsed NMR method. First, when an individual atomic dipole aligns itself with external magnetic field lines, the two are not perfectly parallel. The magnetic dipole of a nucleus precesses slower in a weaker field and faster in a stronger one. The frequency of this precession is called the Larmor frequency, which is exactly the same as the resonance frequency for a given nucleus. So, we can calculate the Larmor frequency for a given proton using the same equation, relating it to the gyromagnetic ratio and effective field.

- Second, in a large collection of similar nuclei, the vector sum of all dipoles within the system will in fact be aligned with the applied field because of the dispersion of the rotating dipoles and the slight excess of the alpha spin state in the group. It is this net magnetization that is the source of the pulsed NMR signal.

X-Ray Crystallography

- In 1895, a new class of electromagnetic radiation was discovered by German physicist Wilhelm Röntgen, who immediately realized the significance of his discovery and published his findings that same year, referring to the uncharacterized radiation as "X-rays." He was awarded a Nobel Prize in Physics—the first ever—in 1901, just six years after his discovery.

- But it would be a decade after Röntgen's Nobel Prize that one of the most impactful applications of his X-rays would be developed. English physicist Sir William Lawrence Bragg, working in collaboration with his father, opened up this entirely new part of the electromagnetic spectrum to the art of identifying organic molecules when he developed a technique to elucidate the structure of crystalline materials using X-rays.

- Bragg and his father had recognized that crystalline materials had a peculiar effect on X-rays as they interacted. There is a scattering effect caused by the electron cloud of each atom in the solid substance, producing a spherical, diffracted wave of the same wavelength. In a crystalline material, the atoms or ions are spaced regularly and evenly, creating planes of evenly spaced positions capable of causing these spherical waves to form.

© stockdevil/iStock/Thinkstock.

The discovery of X-rays earned Wilhelm Röntgen the first Nobel Prize in Physics and gave us an invaluable diagnostic tool.

- In 1913, Bragg and his father famously published a paper detailing the crystal structure of diamond by X-ray diffraction. This structure proved for the first time that carbon was in fact a tetrahedral atom in its sp^3-hybridized state, as chemists had suspected for decades. For their contribution, Bragg and his father shared the Nobel Prize in Physics in 1915.

- X-ray diffraction has since been used to validate many structural theories, including some of those offered by chemists long before its creation. It validated Kekule's benzene ring with Robert Robinson's six evenly spaced carbon atoms about a symmetrical six-membered ring. It also has been used to test Hückel's rule for aromaticity and anti-aromaticity using cyclooctatetraene, which clearly has a ring pucker and also two different carbon-carbon bond lengths within the ring.

- In more recent years, it has helped to produce many amazing protein and DNA models, including the remarkable potassium channel protein structures that won Roderick MacKinnon his Nobel Prize. In short, X-ray crystallography may be the most powerful structural determination tool ever invented.

Mass Spectrometry

- Mass spectrometers were first developed to help us better characterize the structure of atoms, but modern mass spectrometers are used commonly in a variety of applications, including sequencing sophisticated proteins, characterizing DNA, and learning more about complex biomolecular systems.

- The influential analytical technique was first developed by British scientist Sir Joseph John Thompson in the first decade of the 20th century. Thompson won the 1906 Nobel Prize for his discovery of the electron a decade earlier.

- As he continued his research into subatomic particles, he turned his attention to the nucleus of the atom, seeking any clue of the existence of subatomic particles there. He found what he was looking for when he was able to create a beam of ionized neon gas in his lab. He aimed his beam at a piece of photographic paper, passing it through a magnetic field to get there.

- Thompson knew that when a charged particle moves through a magnetic field, it experiences a force that is dependent on its charge and velocity, but not its mass. So, basically, the field pushes all ions moving through it with equal force. Naturally, he expected to see all of his neon ions deflected to the same extent, striking the photographic paper in the same spot.

- What he saw was remarkable: Two different spots formed on the developed paper. Thompson had observed neon atoms behaving as if they had two different masses. He had proven the existence of elemental isotopes—atoms of the same element with different molecular masses. The only logical explanation for his observation was that atomic nuclei are made up of something smaller than themselves.

- But mass spectrometry is useful for much more than the study of atomic structure. In the past half century, it has become a favorite technique of chemical researchers and forensics laboratories because of its amazing ability to identify unknown compounds quickly and efficiently.

- There are three main components to a mass spectrometer: an ionizer, a mass analyzer, and a detector. As the name implies, the ionizer is responsible for creating a beam of gas-phase ions. These ions then pass through a mass analyzer, which modifies their motion using an applied magnetic field. Finally, the ions collide with a detector placed such that only ions taking a specific trajectory will strike it.

- How is it that mass spectrometry is so useful for identification when so many molecules share the same molecular mass? The answer lies in a phenomenon that takes place during the ionization process.

- During the high-energy process of ionization, many of the ionized organic compounds fragment into smaller pieces. Usually, this involves the homolytic cleavage of a bond to form two pieces of the original molecule—one neutral radical and one cation.

- The mass spectrum not only gives us the mass of the molecule by detecting its molecular ion, but also produces a sophisticated fingerprint of the molecule through a process called fragmentation. By measuring the mass difference between the molecular ion and its fragments, we can determine the identity of the fragment that was lost to produce each fragment. Not the molecular mass, but the fragmentation pattern is the key to much of the utility of this technique.

The Limitation of Spectrometry

- We have all probably heard the amusing story of a person making the unfortunate decision to eat a poppy seed bagel before a urinalysis drug screening only to be told afterward that he or she has tested positive for heroin use. This situation has been explored by popular programs like *MythBusters* and has been parodied on popular comedy programs like *Seinfeld*.

- If the mass spectrometry used in the urinalysis is so powerful and accurate, then how can it be defeated by something as simple as a poppy seed? The answer is a perfect illustration of the limitation of spectroscopy and spectrometry in general.

- Morphine and heroin only differ in structure by the presence of two acetyl groups. These acetyl groups are added to the naturally occurring morphine from poppy seeds by esterification with acetic anhydride. The seed from mature flowers is needed.

- Once injected, a fraction of the heroin molecules are slowly hydrolyzed back into morphine by enzymes in the blood. The resulting morphine is modified and further broken down in the liver. Finally, these metabolites are eliminated from the user's system, ending up in their urine.

© Junghee Choi/iStock/Thinkstock.

The metabolic by-products of heroin are practically identical to those of morphine, which are present naturally in poppy seeds.

- Many drug urinalysis techniques use mass spectrometry to search for the metabolic by-products of heroin, which are practically identical to those of morphine, which are present naturally in poppy seeds used in the food industry.

- This is a fantastic example of how the utility of even the most powerful instruments is not only a function of the technology, but also of the human element. In short, no spectral analysis is ever complete without careful contemplation on the part of the researcher.

Suggested Reading

McMurry, *Fundamentals of Organic Chemistry*, 13.1.

Questions to Consider

1. What are some non-spectroscopic applications of the Fourier transform?

2. How does changing the wavelength of diffracted light change the diffraction pattern for a given crystal?

3. How is the fragmentation pattern of a compound used to identify it in a mass spectrometry experiment?

Advanced Spectroscopic Techniques
Lecture 31—Transcript

Over our last few lectures, we've considered some of the techniques that organic chemists use to characterize organic compounds and their structures. Specifically, we've looked at UV-visible spectroscopy, infrared spectroscopy, polar imagery, and continuous-wave NMR spectroscopy. Now, all of these techniques were developed in the late 1800s and early 1900s and have no doubt contributed significantly to developing the basis for our understanding of organic chemistry. But in many cases, most notably the case of NMR, these techniques are not locked in time. They've been relentlessly expanded, improved upon, and even supplemented with newer techniques for identifications, just a few of which we're going to explore today.

Now, last time we explored the technique pioneered by Isidore Rabi and his contemporaries in the middle of the 20th century. Specifically, we discussed the phenomenon of resonance and how factors like shielding and magnetic coupling can all affect the resonance behavior of specific protons within a sample.

Today, we will apply all of these principles and more as we investigate modern NMR spectrometers, which can not only collect the spectra faster and more precisely, but are fundamentally different in the way which they do so. Several advances in technology have led to the development of what's called the pulsed NMR technique. First and foremost is the development of superconductors. See, all modern high-field spectrometers use a superconducting magnet to generate an extremely high applied field, which helps us to achieve better resolution in the spectrum.

Keeping these superconducting magnets cold is critical to their function. So spectrometers contain a series of concentric vacuum-walled compartments called Dewars. There's an exterior Dewar, filled with liquid nitrogen, and an interior Dewar, containing liquid helium and the superconducting magnet itself. The liquid helium keeps the magnet itself at four Kelvins. That's colder than the dark side of the moon. Meanwhile, just inches away from it, is the organic sample at room temperature.

The second advancement which opened the door to the pulsed NMR method is computer technology. The detector in NMR is more often called a receiver, because the response it creates is not a spectrum, but rather a raw data set which has to be manipulated by a computer before it can be used. So, these two technologies, the superconducting material development and computing power advancement, have allowed us to use this very effective technique to do so much more with NMR than could ever have been done in Rabi's day.

As we try to explain the pulsed NMR method, we're going to rely on all the knowledge that we've gained in the previous lecture, that in the presence of a magnetic field, protons align their dipoles in the parallel alpha-spin state or the anti-parallel beta-spin state, and that the populations of these two states is governed by the Zeeman splitting. Zeeman splitting is the energy difference between the alpha and beta states and can be calculated using the field strength at the nucleus and the gyromagnetic ratio.

But to really understand pulsed NMR will require us to acknowledge two additional principles. First, when an individual atomic dipole aligns itself with external magnetic field lines, the two are not actually perfectly parallel. You see, the dipole precesses about the applied field, much like this gyroscope, which I'm about to wind up so that we can see this for ourselves. So you can clearly see what's going on here. Let me quiet it down a bit for you so we can discuss. So the axis of that gyroscope was precessing about the axis of the Earth's gravitational field, which is oriented straight up and down. Now, just as that gyroscope would precess, say, slower in the reduced gravity of the moon but faster in the higher gravity of Jupiter, so will the magnetic dipole in the nucleus precess slower in a weaker field and faster in the stronger one. The frequency of this procession is known as the Larmor frequency, which is exactly the same as the resonance frequency for a given nucleus. So we can calculate the Larmor frequency for a given proton using the same equation relating it to the gyromagnetic ratio and effective field.

Our second consideration, in a large collection of similar nuclei, the vector sum of all dipoles within the system will, in fact, be aligned with the applied field, because of the dispersion of the rotating dipoles and the slight excess of the alpha spin state in the group. And it's the net magnetization which is the source of the pulsed NMR signal.

So let's set the stage for our first pulsed NMR experiment. First we need to get our bearings. So let's define a coordinate system such that three axes, x, y, and z, are present. The z-axis will be aligned with the applied magnetic field lines, and therefore, also with the net magnetization of the sample. Now, what makes pulsed NMR so fundamentally different from the continuous-wave experiment is this, there's no continuous wave. There is simply the magnetization generated by the protons within the sample. So, they rely on a hard pulse of radio energy containing a very broad range of frequencies. This type of pulse, electromagnetically pushes the net magnetization downward into the x-y plane.

Now, in my example, I've place the magnetization along the x-axis. But just as soon as that hard radio frequency pulse is removed, the magnetization wants to return back to its original orientation. But here's the catch, it's path is not straight. Instead, it precesses about that z-axis at its Larmor frequency as it relaxes. It's this precession which is the basis for the raw data collected in a pulsed NMR experiment.

But just how is this frequency measured? Well, consider first the analogy of the lighthouse with a beacon of light rotating at a given rate. If an observer stands at a fixed location and watches that lighthouse beacon rotate over a given period of time, what will they observe? Well, they'll observe the intensity of the beam first diminishing as the source rotates away, and then intensifying, as it sweeps back in the observer's direction. And this process will continue for as long as that beam rotates. If we were to plot this intensity as a function of time, we would expect to observe a trigonometric function, specifically, a cosine wave. So a Fourier transform of this function should transform my cosine function into the time domain into a single peak in the frequency domain. Modern NMR spectrometer works on a similar principle, listening with the radio antenna, which produces an alternating current in response to the rotation of the magnetic vectors in the sample.

Now, where this analogy breaks down is the fact that the magnetization of our sample not only rotates, but also relaxes back to its original orientation. So, the x-y vector component of this magnetization is not only rotating, but also weakening as the experiment progresses. And this produces a decaying AC current at the detector. It's a plot of that decaying AC current intensity versus time, which is the raw data produced by the spectrometer. And we call this the free induction decay, or FID for short.

So now that we understand how a spectrometer collects information about the Larmor frequency of a collection of protons, I want to show you how we convert this information into something more useful to us, like a completed NMR spectrum, the likes of which we saw last time in the continuous-wave experiment. So let's begin by looking at a raw data set for a pulsed NMR experiment, known as a free induction decay. On my horizontal axis will be time because I'm in the time domain. And my vertical axis will be the intensity of the AC current. So I expect to see something like this.

Now, what's interesting about this particular data set is that it's so simple, that I can interpret it without transforming it. So let's do that first. There are really only two different frequencies embedded within this particular free-induction decay. One frequency which is rather fast, frequency A, so the amount of time for one oscillation to take place here is very short. But if you look carefully, frequency A is embedded on top of a larger, longer frequency here. This would be frequency B. So my data set contained two frequencies. Now, NMR data sets for actual samples can contain dozens or even hundreds of different frequencies, all superimposed on one another. So it's not always quite this easy.

And to solve that problem, we use a mathematical operator known as the Fourier transform. The Fourier transform is very complex, and we won't cover in detail today exactly how it's done. But what's important here is the outcome. If I Fourier-transform my data set, I alter the time domain and start looking at my data instead in the frequency domain. So, my higher frequency, frequency A, is somewhere down here in the downfield region of my continuous-wave experiment, whereas a longer frequency, or a lower frequency, like frequency B, would be somewhere up here in my upfield region of the spectrum.

So we actually plot our data with higher Larmor frequencies at the left-hand side of our plot, which might seem a little bit counterintuitive. But remember, this is the same type of information that's included in a continuous-wave experiment. So, to be sure that we can compare spectra collected in, say, the year 2014 with spectra collected in the 1950s, we alter the axis in such a way that our frequencies are sort of reversed from what we would expect.

So let's move on to another technique. In 1895, a new class of electromagnetic radiation was discovered by German physicist Wilhelm Röntgen. Röntgen immediately realized the significance of his discovery and published his findings the same year, referring to the uncharacterized radiation only as x-rays. Now, Röntgen was not the only one who saw potential in his new discovery, either. He was awarded the Nobel Prize in physics. In fact, the first one ever, in 1901, just six years after his discovery.

But it would be a decade after Röntgen's Nobel Prize Award that one of the most impactful applications of x-rays would be developed. English physicist William Lawrence Bragg, working in collaboration with his father, opened up this entirely new part of the electromagnetic spectrum to the art of identifying organic molecules when he developed a technique to elucidate the structure of crystalline materials using x-rays. Bragg and his father had recognized that crystalline materials had a peculiar effect on x-rays as they interacted. There's a scattering effect caused by the electron cloud of each atom in the solid substance, producing a spherical diffracted wave of the same wave length as the one which first came into the crystal.

Now, in a crystalline material, the atoms or ions are spaced regularly and evenly, creating planes in evenly spaced positions capable of creating these spherical waves which propagate out. Now, to give you an idea of how we can use this to get information about structure, I'm going to show you a very simple relationship between just two atoms. Each atom is represented here by one of my blue spheres. So in the left case example, my two atoms are closer together in space than are the two on the right side.

So, how can I use x-rays to get information about the distances between these two? Well, the key is in the scattering. So if I have a coherent X-ray beam, which, in this case, is going to be coming in from the left-hand side of my monitor and striking these atoms, their electron clouds respond by emitting a spherical wave. And that spherical wave has a very well defined wavelength, exactly the same as the incident light. So, let's create, in this case, since it's on a two dimensional surface, circles, but they'll represent the spherical waves that are created during diffraction.

The pattern looks pretty complicated, doesn't it? But, it's actually not as complicated as you may think. Now, x-rays have the ability to expose photographic film. So if I place a photographic plate in a certain proximity to my crystal, only those locations where the propagating spherical waves interfere with one another constructively should I see exposure. So, in this case, I expect to see a pattern of spots developing along my film, where there is constructive interference causing those spherical ways to continue propagating through space.

Now let's do the same experiment but on my atoms that are spaced farther apart. Again, my spherical waves propagate and in some places interfere constructively and in others destructively, but if I place my photographic plate at the same distance and do the same analysis, look what's happened. The spatial relationships of the exposures on my film have changed. So by analyzing the patterns that result from this diffraction phenomenon, we can back calculate the spatial arrangement of atoms within a crystalline structure. This is the principle on which X-ray diffraction works.

Now understanding how the diffraction patterns develop and actually observing them is always challenging for students, in part because our eyes can't see x-rays. But our eyes can see visible light. And while the distances between atoms are so small that they require x-rays to resolve, we can create little gadgets that we use in the classroom in the laboratories to show a diffraction pattern using visible light. I'm going to do that for you right now, so to help you out, I need to dim the lights.

All right, I'm going to begin by showing you a dot from my laser pointer. Not very exciting, is it? But it's coherent light; it's light of a single wavelength. And when that light interacts with the grating, which is spaced somewhere around the wavelength of that light, I can cause that light to diffract, just like the x-rays do within a crystal. For example, this is a diffraction pattern for a relatively simple grate. And depending upon the spacing and the geometry of the separations, the striations in the slides I'm using, I can create different patterns using just one laser pointer. And encoded within these patterns is all the information about the geometry within the striations in the slides. In fact, with a little bit of practice and the right kit, I can show you the exact structure that Rosalind Franklin, Watson, and Crick saw when they characterized the structure of DNA for the very first time. This, in fact, contains all the symmetry elements of a DNA double helix.

It's a pretty amazing trick, isn't it? All of those different patterns created using just a single beam of coherent red light and some special slides, which, you can't quite tell by looking at them, but they contain striations that are just a few hundred nanometers wide, and therefore, interact with that visible light in a very similar fashion to the way that the atoms within a crystal interact with x-rays. So by shining a beam of x-rays on a crystal and exposing the photographic film to the diffracted waves, a circular pattern like those we just saw with visible light can be obtained. Encoded within this image are all the angles of these symmetry planes within the crystal that produced them.

In 1913, Bragg and his father famously published a paper detailing the crystal structure of diamond by X-ray diffraction. This structure is special because it proved for the first time that carbon was in fact a tetrahedral atom in its sp^3 hybridized state, just as chemists has suspected for decades. And for that contribution, Bragg and his father shared the Nobel Prize for Physics in 1915.

X-ray diffraction has since been used to validate many structural theories, including some of those offered by chemists long before its creation. A great example of this is the concept of aromaticity and antiaromaticity. It validated Kekule's benzene ring with Robert Robinson's six evenly spaced carbon atoms about a symmetrical six-membered ring. It also has been used to test Huckel's rule for aromaticity and antiaromaticity using cyclooctatetraene, which clearly has a pucker to the ring and also clearly has two different carbon-carbon bond lengths within the ring because of its antiaromaticity. In recent years, it has helped to produce many of the amazing protein and DNA models which we used in the associated units of this course, including the remarkable potassium-channel protein structures, which won Roderick MacKinnon his Nobel Prize. In short, X-ray crystallography may be the most powerful structural determination tool ever invented.

Our final technique in this lecture will be mass spectrometry. Now, we call this technique spectrometry rather than spectroscopy, because, unlike the other techniques that we've discussed so far, in mass spectrometry we're not measuring the direct interaction of light with matter. Mass spectrometers were first developed to help us better characterize the structure of atoms. But modern mass spectrometers are used more commonly in a variety of applications. They find use in criminal forensics labs, searching for the residue of gunpowder or the metabolites of illegal drugs. Anyone who's traveled through the American airport system in recent decades has almost certainly seen a security agent swab the baggage or clothing of a traveler, inserting that swab into a small device, and then waiting for result before allowing that passenger to pass on. Well, the device that they're using is a mass spectrometer, searching for trace amounts of chemicals used in the creation of explosives. Modern mass spectrometers even have the ability to help us sequence sophisticated proteins, characterize DNA, and learn more about complex biomolecular systems too.

The influential analytical technique was first developed by British scientist Sir Joseph John Thompson in the first decade of the 20th century. Thompson was already a prolific physics researcher, having won the 1906 Nobel Prize for his discovery of the electron a decade earlier. Now, you might think that reaching such a pinnacle of scientific achievement would mean that Thompson would relax and enjoy his acclaim. But most researchers with the passion, talent, and determination required to win the Nobel Prize are not the kind to rest on even such a high laurel. And Thompson was no exception.

As he continued his research into subatomic particles, Thompson turned his attention to the nucleus of the atom, seeking any clue of the existence of subatomic particles there. He found what he was looking for when he was able to create a beam of ionized neon gas in his lab. He aimed that beam at a piece of photographic paper, passing it through a magnetic field to get it there. And Thompson knew that when a charged particle moves through a magnetic field, it experiences a force which is dependent on its charge and velocity, but not on its mass. So basically, the field pushes all ions moving through it with equal force. So naturally, he expected to see all of his neon ions being deflected to the same extent, striking the photographic paper in the same exact spot.

But what he saw was different and remarkable. Two different spots formed on the developed paper. So, Thomson had observed neon atoms behaving as if they had two different masses. He had proven the existence of elemental isotopes—atoms of the same element, different atomic masses. The only logical explanation for his observation was that atomic nuclei must be made up of something smaller than themselves.

But mass spectrometry is useful for much more than the study of atomic structure. In the past half century it's become a favorite technique of chemical researchers and forensics laboratories because of it's amazing ability to identify unknown compounds quickly and efficiently. So let's get busy exploring a simple mass spectrometry experiment using an organic compound.

For this discussion, I'll be using a rather simple molecule, pentane, and describing the most basic of mass spectrometry techniques. The mass spectrometry experiment is extremely sensitive and requires that we generate a high energy, very reactive species. So the experiment has to take place in a really high vacuum. This basically ensures that frequent collisions of our analyte ions with others or with atmospheric gases are avoided.

Now, inside of a typical mass spectrometer, the pressure is about the same as it is on the outside of the International Space Station. There are three main components to a mass spectrometer, an ionizer, a mass analyzer, and a detector. As the name implies, the ionizer is responsible for creating a beam of gas-phase ions. These ions then pass through a mass analyzer, which modifies their motion using an applied magnetic field. Finally, the ions collide with the detector, placed such that only ions taking a specific trajectory will strike it.

So, step one for our pentane molecule is the ionization process. In our example, we're going to use electron impact ionization as the ionization technique. This method involves passing the gas-phase pentane molecules through a beam of high-energy electrons. And, as this happens, occasionally an electron strikes a pentane molecule, knocking an electron out of its cloud. Now, this process generates a radical cation of pentane. Something like this would most likely never form in a lab bench. But in a high vacuum, being bombarded with electrons, we can expect to see some pretty strange things.

If the pentane survives the ionization process, it moves to a magnetic field, which is carefully shaped to deflect ions in a known direction. Positively charged ions, like ours, moving through a field, will experience a force. Since force equals mass times acceleration, the acceleration, or curvature, of the ion's path is inversely proportional to its mass. And because all moving, singly-charged positive ions experience the same force in our field, we can say that more massive ions are deflected to lesser extents. So, knowing the mathematics governing the physics of deflection, we simply increase or decrease the strength of the magnetic field until the detector indicates a hit. So, when using only monovalent ions, meaning ions with a single charge, we expect to see just one peak for pentane and a charge-to-mass ratio of 72. There will also be a few smaller peaks produced by the natural abundance of deuterium and carbon 13 in our sample, and so those will weigh 73, maybe even 74.

But this mass spectrum wouldn't be as useful as I led you to believe, would it? After all, how many different organic compounds are there with a mass of 72? There's plenty. Just for a start, how about the three geometric isomers of *n*-pentane, isopentane, and neopentane? So, how is it that mass spectrometry is so useful for identification when so many molecules share the same molecular mass? The answer lies in a phenomenon we initially neglected, which takes place during the ionization process. So let's rewind and look at our experiment one more time.

So, let's set a few tenets before we continue our discussion here. All mass spectrometry experiments require some kind of ionization. Now, the type that I'll talk about today isn't the only kind, but it's called electron impact ionization, and it's very popular for use in small organic molecules. And in this technique, a high-energy electron is going to impact a molecule and knock another electron out of the cloud in a sort of game of molecular billiards. So, as a cartoon representation, here I have a gas-phase molecule.

Now, my gas-phase molecule has an electron cloud. And I'm going to strike one of the electrons from that cloud with a high-energy electron from outside the molecule, not pleasant for the molecule. But what it does is relieve it of one electron generating a cation with an unpaired electron, which is a very unstable species. So they don't always survive. In fact, they rarely survive. In many cases, they break apart. And when they do so, most often, the charge and the radical, or the unpaired electron, go separate ways.

OK, so let's get out of the cartoon world here and look at a real molecule, pentane. One more time, we're going to ionize this with electron impact. So my pentane has an electron cloud, which is depicted here in motion, and I have an incoming high-energy electron, which strikes one and then ejects it from the electron cloud. I've created a radical cation. But remember, radical cations are very unstable. And in the case of pentane, there can be a fragmentation. I'm going to draw my fragmentation occurring here at one of the terminal carbon-carbon bonds. What takes place most often is a homolytic bond cleavage, which means the two electrons go in different directions. And when this happens, the bond breaks. But not only that, but my radical and my positive charge go to separate sides of that broken molecule, or that broken ion. So in this example, I've created a methyl radical and a butyl cation. And the relative stabilities of these ions and radicals is what's going to dictate how many of them form in my experiment.

So let's take a quick moment here and finish out our day by looking at what we expect to be the mass spectrum of pentane. The mass of pentane, or m over z, as we reported, because z is a factor of charge. But remember, our process only creates charges of one. So my m over z can be thought of as a mass. I have my molecular ion with a mass of 72. But I also can expect there to potentially be fragments involved in the breaking of the terminal bond. When this happens, I can get two possible pairs, a methyl radical and primary cation, or, a primary radical and a methyl cation. In these cases, the masses of the ions generated are 57 and 15, respectively. Similarly, I could have the internal carbon-carbon bond break, generating another two pairs.

But notice that this time, the two pairs that are generated are both primaries, or primary radical and a primary cation, slightly more stable than the others. So my peaks at m over z of 43 and 29 are expected to be a little bit larger. In fact, that's exactly what we see in the mass spectrum of pentane using this method. We see a mass spectrum in which the peaks at 29 and 43 are larger than those at 15, 57, and even at 72. We also see a little bit of those isotope peaks in there. So here we have smaller peaks for the ions that are generated along with radicals that are of lower substitution. We see larger peaks when there's greater substitution. And again, in this case, the molecular ion itself is one of the smallest peaks in the spectrum. So if we continue thinking about our simple example of pentane by comparing its fragments in their abundance with those produced an isopentane and neopentane, we can find clear and predictable differences, which enable us to discriminate between their mass spectrum.

Of course, hydrocarbons are not the only compounds with this molecular mass. Consider acrylic acid, also 72 atomic mass units. Its molecular ion is the same mass as that of pentane. But look at its fragmentation pattern. It's quite different, isn't it? So it's clear from the masses of the fragments that ions created by loss of a hydroxyl group, loss of a carboxylic acid group, and also by loss of a vinyl group are all present. So, this example shows very clearly how not the molecular mass, but the fragmentation pattern, is the key to much of the utility of this technique.

We've all probably heard the amusing story of a person making the unfortunate decision to eat a poppy seed bagel before urine analysis drug screening only to be told afterward they tested positive for heroin use. The situation has been explored by such popular programs as *MythBusters* and parodied on popular comedy programs like *Seinfeld*. If the MSUs in the urinalysis is so powerful and accurate, how then can it be defeated by something as simple as a poppy seed? Well, the answer is actually quite simple. And it's a perfect illustration of the limitation of spectroscopy and spectrometry in general.

You see, morphine and heroin only differ in structure by the presence of two acetyl groups. Now, these acetyl groups are added to the naturally occurring morphine from poppy seeds by esterification with acetic anhydride, a technique which we explored in our esters module. Now, of course, the seed from mature flowers is needed. And I'm leaving out a few important details from the process in the hopes of avoiding a visit from the United States DEA after this taping.

But where the story comes to bear on the limitations of spectrometry is this. Once injected, a fraction of the heroin molecules are slowly hydrolyzed back into morphine by enzymes in the user's blood. The resulting morphine is modified and further broken down in the liver. And finally, these metabolites are eliminated from the user's system, winding up in their urine. Many drug urine analysis techniques use mass spectrometry to search for the metabolic by-products of heroin, which are practically identical to those of morphine, which is present naturally in poppy seeds used in the food industry.

This is a fantastic example of the utility of even the most powerful instruments can be not only a function of the technology, but of the human element, as well. In short, no spectral analysis is ever complete without careful contemplation on the part of the researcher.

So let's summarize the last of our spectroscopic techniques. Today, we looked at pulsed NMR and how modern technology allows us to use superconducting magnets and radio receivers to measure the Larmor frequency of spin one-half nuclei. This reproduces the information from a continuous-wave experiment but in a minuscule fraction of the time.

We touched on X-ray crystallography and how scattering of these high-energy rays by the electron clouds of highly ordered atoms leads to a pattern of interference in the scattered light. And this pattern can be used to back calculate that regular arrangement. We viewed a few of these structures and discussed how this technique has been used to validate many of the structural theories which preceded it.

Then we took a look at mass spectrometry, and we saw how ionization, mass analysis, and detection systems work, and specifically, how electron impact generates the molecular ions and fragment ions, how a magnetic field is used to deflect each ion to a detector, which generates an electrical current when struck by ions. We discussed how the identity and abundance of fragments in the spectrum can be used to deduce the structure of compounds analyzed, even when multiple candidates have the same molecular mass. And finally, we had an object lesson in the human element of identifications and how we must always remember that these techniques have their limitations.

So we finished our short survey of techniques used for identifying organic compounds once they've been made or extracted in the labs. But we have yet to really discuss how we can achieve high levels of purity, the levels of which are necessary to use some of these spectroscopic techniques. We're going to broach this subject in our next lecture when we start by discussing how to get the purest possible sample using a technique known as recrystallization. I'll see you for that talk next time.

Purifying by Recrystallization
Lecture 32

I t is very rare that any organic synthesis runs cleanly to absolute completion, producing no by-products and leaving no starting materials behind. It is even rarer that we find organic materials in highly purified forms in natural sources. So, how do we sufficiently purify organic material? In this segment of the course, you will investigate a few staple techniques used for isolation of organics in modern laboratory and manufacturing settings. Specifically, in this lecture, you will learn about two very important and widely used techniques—one for purification and the other used to determine purity. Both of these techniques rely on the tendency of organic molecules to form highly ordered crystals.

The First Recrystallization Experiment
- To find one of the first examples of mass-scale organic purification, we go to India in the 5th century B.C. Evidence shows that it was around this time that the people of India began harvesting a wild reed that, when they chewed on it, became sweet tasting and pleasant.

- This commodity quickly caught on in other regions of Asia and Europe. Greek physician Dioscorides wrote of it in the first century A.D., and records from the Tang dynasty confirm that it was being cultivated in China as early as 600 A.D.

- But the real turning point in the global spread of this product took place in the 3rd century A.D. During this time, most of what is now northern India was ruled by the Gupta dynasty, which is famous for its deep and altruistic value of cultural development.

- It was during this golden age of Indian culture that techniques were developed to extract the juice from those reeds, and then allow it to form crystals by drying it slowly in sunlight. These crystals, it turns out, retained the sweet flavor of the reeds, yet spoiled far more slowly in their crystalline form.

- They were creating crystalline sugar—sucrose extracted from sugar cane as a solution and then crystallized. By developing this process, the people of the Gupta kingdom were unwittingly performing what may have been mankind's first recrystallization experiment.

Crystalline sugar is extracted from sugar cane as a solution and then crystallized to create a more viable product.

Crystals: Stability and Purity

- It is actually not so surprising that crystalline organics might be some of the very first organic compounds isolated by man. The reason for this is hidden in the Gibbs free energy of crystals. Crystals are highly ordered, repeating arrangements of atoms or molecules.

- But organic molecules can also achieve these highly repeating patterns in the solid state. When they do so, the intermolecular attractions between and among molecules become stronger, and the free energy of the system naturally decreases. Because lower-energy states are favored in nature, highly ordered materials can form if the energetic benefit of their interactions exceeds the entropic penalty of assuming that ordered state.

- If we can make crystals grow very slowly, under conditions favoring the most stable crystal formation with all the strongest intermolecular attractions, we can produce a crystal of high purity. This is a good thing, because we rarely find compounds like sucrose all alone in nature.

- In biologically sourced solutions like cane juice, which is up to 20% sucrose, we find smaller concentrations of its subunits glucose and fructose as well as small amounts of many other biological molecules.

- In the days of the Gupta dynasty, this was not likely to be a concern to them. Crude, solidified, or crystallized sugar was enough to get the job done, but today's chemists and chefs find a need for highly purified sucrose so that they can use it under the more controlled conditions demanded by the state of their art.

The Process of Recrystallization

- Early sugar was most likely produced by evaporating water from cane juice as rapidly as possible. If, however, the solution expelled from sugar cane is slowly allowed to concentrate past the point of saturation, the sugars contained therein will be compelled to take on a solid form or precipitate from the solution.

- If we produce a concentrated solution of organic molecules like sucrose and then somehow induce the system to become supersaturated, meaning that it has more dissolved sucrose than it could normally hold, we expect the growing crystals of sucrose to exclude the other contaminant solutes.

- Let's use a compound called phthalic acid to better understand the process of crystallization. We start with a sample of phthalic acid and a beaker of water in which we can dissolve the phthalic acid. Phthalic acid has a solubility of about 18 g/mL in boiling water but of about 0.6 g/mL in cold water.

- If the phthalic acid isn't quite pure, its crystals aren't as stable as they could possibly be. But how are they going to get to that more purified, more stable state? Clearly, the molecules in the crystal are not going to spontaneously rearrange in the solid phase, expelling the contaminants, so we want to speed up the process.

- We need to disassemble the crystals at the molecular level, separating them all in such a way that the phthalic acid molecules get a second chance to order themselves in a more stable state. One way to accomplish this separation is to dissolve the crystals, separating them all with solvent.

- We can create a saturated solution of phthalic acid in boiling water. At the beginning, the phthalic acid has no incentive to precipitate, because it is happy staying dissolved in the hot solution. But then, we turn off the heat source.

- As the solution cools slowly, the solubility of the phthalic acid drops dramatically. Then, the organics have nowhere to go but out of solution. The key here is that we are allowing the process to happen slowly. By cooling slowly, we avoid the trapping of impure molecules in the growing crystals.

- At the completion of the recrystallization, the large, well-formed, pure phthalic acid crystals can simply be removed from solution using filtration. The solution from which they came, sometimes referred to as the mother liquor, contains the impurities still in solution—and, unfortunately, a small amount of the phthalic acid. But it's a small price to pay for such a high-purity product, and in this case, it is about $1/30^{th}$ of the total amount of phthalic acid in the sample.

Melting: Determining Purity of Crystalline Solids
- Pure crystals tend to be more stable than their impure counterparts. This phenomenon can be used to purify a compound by crystallizing it from a solution, but we can also exploit this phenomenon in a completely different way—by observing the melting behavior of a sample to help us identify it and assess its purity.

- Just as dissolution involves liberating individual molecules from one another by solvation, melting liberates them by heating them until they have sufficient kinetic energy to overcome those forces holding the solid together.

- With this in mind, the logic behind using melting to assess purity becomes clear—that more-purified crystals have greater forces holding them together, meaning that they require more energy to melt. In other words, as purity increases, so does the melting point.

- This concept can be illustrated using a hypothetical phase diagram for a binary system. In order to construct this phase diagram, we take the diagram for a pure substance, which we will call substance A. Its phase diagram includes all of the phase transitions, complete with boundaries for melting and freezing, boiling and condensing, and sublimation and deposition.

A Typical Binary Phase Diagram

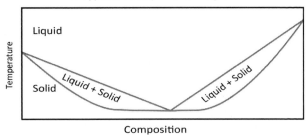

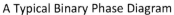

- However, the diagram only applies to a pure substance. To create a binary phase diagram, we settle on a constant pressure like 1 atmosphere, and at that pressure, we project the diagram along a third axis of composition. We do this for two hypothetical solids, which we will call A and B, focusing on the melting transition.

- The new x-axis is the mole percentage of one of the two compounds, so the extremes of the axis correspond to pure materials, and the interior corresponds to mixtures of the two in various ratios. The melting process for most mixtures occurs over a range of temperatures rather than a single, distinct temperature.

- The lower temperature barrier for melting is called the solidus. At temperatures below the solidus for a given mixture, no liquid can coexist at equilibrium. The higher temperature barrier is called the liquidus, and it represents the temperature above which only liquid can exist at equilibrium.

- This diagram will allow us to predict the melting behavior of our system as a function of its composition at 1 atmosphere of pressure. The melting points of pure A and B are just that—points of a single temperature at which the solidus and liquidus converge. After all, we learn in general chemistry that the melting of a specific pure substance at 1 atmosphere pressure takes place at one specific temperature.

- But let's consider what happens when we mix a bit of compound B into the solid sample of A—just about 5 mole percent. At this composition, the presence of a few molecules of impurity B are disrupting the crystals of A, reducing its melting point.

- But there is even more going on here. The temperatures at which solid and liquid can coexist are now a range, rather than a single temperature. This is because a process called incongruent melting is taking place.

- Incongruent melting means that the liquid generated first is not of the same composition as the solid. So, the liquid is more concentrated in B than in the solid, meaning that the melting point of the crystal increases as it melts, although it never quite reaches the melting point of the pure solid.

- As we continue to add more and more compound B to the solid and repeat the experiment, we see the trend of melting point depression and broadening continue, until eventually the gap narrows again and comes to a point at a local minimum on the solidus.

- We call this convergence of the solidus and liquidus the eutectic point for the mixture. It is the only temperature at which a mixture of the two components can coexist at one distinct temperature.

- Think of it as the point at which B stops being the impurity in A and the roles are reversed. Adding even more and more B reverses the trend, finally taking us to its own melting point at the far end of the projection.

- This kind of complex phase behavior gives us a very useful tool to confirm the identity and purity of a compound. We use a device called a MelTemp apparatus, which consists of a heating block with a temperature control and a viewing window. If we load a sample of the original crude phthalic acid and one of the recrystallized material, we can confirm that the melting point of the purified material is higher.

- In addition, a third sample consisting of the purified phthalic acid ground thoroughly with a small amount of standard allows us to confirm its identity, because adding anything but the same compound to itself should reduce its melting point. If the sample were anything but phthalic acid, the melting point of the mixture should be depressed.

- We call this simple technique a mixed melting point analysis, and it is a very commonly used quick check for purity and identity when synthesizing compounds in the lab.

Suggested Reading

Wade, *Organic Chemistry*, 3.5B.

Williamson, *Organic Experiments*, Chaps. 3 and 4.

1. Is purification by recrystallization driven by enthalpy or entropy?

2. What are some characteristics of a good recrystallization solvent?

3. How does the inclusion of small amounts of impurity lower the melting point of a given compound?

Purifying by Recrystallization
Lecture 32—Transcript

In our journey through introductory organic chemistry, we've spent a great deal of time looking at ways to prepare organic compounds from one another, as well as ways to characterize the starting materials and products of the reactions that we have undertaken. But we've left one very important question yet unanswered. You see, it's very rare that any organic synthesis runs cleanly to absolute completion, producing no by-products and leaving no starting materials behind. It's even rarer that we find organic materials in highly purified forms in natural sources. So knowing this, we have to ask ourselves the question. How do we purify organic material sufficiently to study the molecules which make it up, or, for that matter, to use as food additives, medicines, or other applications which require very high levels of purity?

In this segment of the course, we will investigate a few staple techniques used for isolation of organics in modern laboratory and manufacturing settings. Today in particular, we'll be interested in taking a look at two very important and widely used techniques, one for purification and the other to determine purities, and both of them rely on the tendency of organic molecules to form highly ordered crystals.

To find one of the first examples of mass-scale organic purification will take us to India in the 5th century B.C. Now, evidence shows that it was just around this time that the people of India began harvesting a wild reed, which, when they chewed on it, became sweet tasting and pleasant.

This commodity quickly caught on in other regions of Asia and Europe. The Greek physician Dioscorides wrote of it in the 1st century A.D., and records from the Tang Dynasty confirm that it was being cultivated in China as early as 600 A.D. But the real turning point in the global spread of this product took place in the 3rd century A.D. During this time, most of what is now northern India was ruled by the Gupta Dynasty. The Gupta Dynasty is famous for its deep and altruistic value of cultural development. And economically beneficial cultural benefit? Well, that was that much better. So it was during this golden age of Indian culture that techniques were developed to extract the juice from these reeds, then, allow it to form crystals by drying it slowly in the sun. These crystals, it turns out, retained the sweet flavor of the reeds, yet, spoiled far more slowly in their new crystalline form.

OK, by now, you probably have put two and two together and figured out that I'm talking about crystalline sugar—sucrose. It's extracted from sugar cane as a solution and then crystallized to create a more viable product. Now, what you may not have known, though, is that by developing this process, the people of the Gupta kingdom were unwittingly performing what may have been mankind's first-ever recrystallization experiment.

So, how is it that crystalline sucrose was stumbled onto so early in human history? After all, when most of us think about crystals, we think about very rare, expensive, and difficult-to-acquire materials, materials like gems and advanced electronics and lasers. Well, it turns out that it is really not so surprising at all that crystalline organics might be some of the very first organic compounds isolated by man. The reason for this is hidden the Gibbs free energy of crystals. Now, crystals are highly ordered. They're very ordered repeating arrangements of atoms or molecules. But organic molecules can also achieve these highly repeating patterns in the solid state. When they do so, the intermolecular attractions between and among the molecules become stronger and the free energy of the system naturally decreases. Since lower energy states are favored in nature, it stands to reason that highly ordered materials conform if the energetic benefit of their interactions exceeds the entropic penalty of assuming a highly ordered state.

So, let's take a look at a crystal structure forming and see if we can get our heads wrapped around this idea a little bit more. What makes pure crystalline materials relatively easy to come by? Let's start with a compound called benzamide. Now, benzamide has a fairly simple crystalline structure. It starts off with a pattern of repeating molecules in one of two orientations, as I've shown them here. You notice the benzene ring in the amide groups pointing either one way or the other in each of the molecules within what would be the surface of a growing crystal.

So here's my surface. Now, as a cartoon, I'm going to just grow my crystal by dropping in the molecules. We'll be talking very soon about how to practically accomplish this. But for now, let's just put our crystal together. So let me drop in a few more benzamide molecules in the proper orientation. OK, now that repeating structure should be even more clear to you. It's very obvious that there are certain regions of the crystal where molecules are oriented in specific directions in an array that arranges them relative to one another in a very, very tight repeating pattern.

This pattern creates a network of intermolecular attractions, including hydrogen bonds. Now, if I highlight them in the crystal here that I've created, it's very easy to see the repeating patterns. In addition to the stabilizing hydrogen bonds, there are dispersion forces, which we would also call London dispersion forces, or van der Waals attractions. And this takes place where the pi systems of the benzene rings stack against one another, very similarly to the way that purine and pyrimidine bases do in DNA and the double helix. So, there's an array of very carefully crafted intermolecular forces holding this crystal together. So this optimized set of intermolecular attractions leads to a very strongly held together, very stable, crystal of this particular organic material. But what if I introduced something else? What if my sample is not absolutely pure? What kind of effect will that have on the energy of this crystal? Let's do that experiment right now.

Let's rebuild our crystal of benzamide first. But this time, I'm going to insert a few impurities. So again, here's my beautiful repeating pattern of my organic compounds aligned perfectly, so I have a beautiful set of hydrogen bonds and van der Waals forces all holding that crystal together. Now, I'm going to take these four molecules and make one very subtle change to them. I'm going to change them from benzamide to benzaldehyde, taking away the NH_2 group. Now look what happens when I do that. I have removed the ability of these molecules to form the hydrogen bonds which stabilized the crystal in this very specific, localized position. So all of these hydrogen bonds that I've colored in red no longer can be formed.

So by including just a few molecules of benzaldehyde in my pure crystal, or otherwise pure crystal, of benzamide, I have decreased the overall strength of the intermolecular forces holding them together. So the fact that I'm missing these intermolecular attractions means that the free energy of my crystal is higher. In other words, it's less stable; it's a weaker crystal, and it's easier to disassemble.

So, if we can make crystals grow very slowly under conditions favoring the most stable crystal formation with all the strongest intermolecular attractions possible, we should be able to produce crystals of very high purity. And this is a good thing, because we rarely find compounds like sucrose all alone in nature. And biologically-sourced solutions, like cane juice, they can be up to 20 percent sucrose. But we find smaller concentrations of all of its subunits, like glucose and fructose, as well as small amounts of many other biological molecules.

Now, in the days of the Gupta Dynasty, this was not likely to be a concern to them. Crude, solidified, or crystallized sugar was enough to get the job done. But today's chemists, and chefs for that matter, find need for highly purified sucrose so that we can use it under more controlled conditions demanded by the state of the art. So, early sugar was most likely produced by evaporating water from cane juice as rapidly as possible. If, however, the solution expelled from sugar cane is slowly allowed to concentrate past the point of saturation, meaning, the sugars contained therein will be compelled to take on a solid form or precipitate from solution, well, if we produce a concentrated solution of organic molecules, like sucrose, then somehow induce that system to become super saturated, meaning that it has more dissolved sucrose than it normally could hold, we expect the growing crystals of sucrose to exclude the other contaminant solutes in order to achieve optimal intermolecular attractions.

To better understand the process of crystallization, let's take a look at a compound called phthalic acid. I have a sample of phthalic acid here on the bench. I also have a beaker of water, which I can use to dissolve the phthalic acid. Now, phthalic acid has a solubility of about 18 grams per milliliter in boiling water, extremely high. But it's only about 0.6 per milliliter in cold water. So if my phthalic acid isn't quite pure, its crystals aren't as stable as they could possibly be. But how are they going to get to that more purified, more stable state? Clearly, the molecules in my crystals are not going to spontaneously rearrange in the solid phase, expelling any contaminants that might be in them, at least not in my lifetime. And this camera crew, they're paid by the hour. So let's see if we can speed it up a little bit, OK?

What we need to do is disassemble these crystals at the molecular level, separating them all in such a way that the phthalic acid molecules get a second chance to order themselves in a more stable state. Of course, one way to accomplish this separation is to dissolve the crystals, separating them all within a solvent. So I've set myself up to do this task right here. What I have is a beaker containing phthalic acid that's been purchased, which is about 98% pure; a smaller beaker which I'm not going to touch—this is phthalic acid that I've contaminated with a red organic molecule so that we can see that contaminant as it behaves in this system that we're about to subject it to. In the larger beaker here is what I'm actually going to be purifying. In order to purify this I'm going to use nothing more than water and some heat. This water's already hot, so let me get my safety glasses on just in case. I'm going to turn my hotplate back on, which is also a magnetic stir plate. It contains a magnetic element underneath of the surface of the ceramic hot plate, which I can use to articulate this little magnetic spin bar inside. This allows me to more thoroughly heat my solvent more rapidly.

While that's going I'm going to put on some gloves. Hopefully they'll stay in one piece while I put them on. These are disposable latex gloves. They're really just designed to protect my hands from the chemicals I'm dealing with, even though they're fairly safe as organic compounds go. Very soon my water will be boiling. I pre-heated it to accelerate this process, so that we can see what goes on. What I'm going to do as soon as this gets warm enough is add my water to this beaker containing phthalic acid and a little bit of the red organic contaminant to see if we can get rid of it by disassembling it at the molecular level through dissolution and then encouraging it to crystallize, but very slowly. Since this is a very hot flask of water I'm going to put a glove over my hand to protect it from the heat as well as the chemicals. Let's get started with the dissolution process. Again, this is fairly hot water but you can clearly see that it wasn't completely boiling. So it's going to require a little bit more help to get it to thoroughly dissolve. In order to do this, I've given myself an extra spin bar which I'll add to the beaker now. I'm going to return everything to my hotplate. We can see that the spin bar is moving because the powder is flowing. Now we need to give it some time to heat up.

In this case I'm heating my recrystallization solution on a hotplate. The surface of the hotplate can reach temperatures well in excess of 300 degrees centigrade, which is hot enough to cause many organic solvents to catch fire spontaneously in air. The only reason that I'm heating this particular solution the way that I am now is because the solvent that I'm using is water, which is not flammable. Water is essentially hydrogen that's already been burned, so it's not possible to burn it any further. In a future lecture, we'll actually take a look at some ways that we do heat organic solvents in safer ways than what I'm doing here, again if this were in fact an organic material.

So we've finished boiling our solution of phthalic acid. Now, as you can see, I'm so saturated that I'm already starting to get a little bit forming on the top of my solution if you've got the right angle for that. And also, you can see that I've had to add a little bit more water as the boiling process went on to be sure that I got the vast majority of my compound dissolved. So now we have to wait and allow slow cooling to take hold. And the reason that slow cooling needs to take place is that we want to give all of the phthalic acid molecules in solution ample time to find those perfect orientations as they come in and become part of the growing crystal. If we do this, any other molecule that tries to get involved with that crystal will not adhere as strongly as phthalic acid and can simply be excluded as the crystal grows.

So, if my original solution was made correctly and I have thoroughly separated molecules of phthalic acid and impurity, I should have only phthalic acid growing back in its crystals, leaving the impurity behind in solution. But that's going to take a little bit of time, because this has to cool all the way from 100°C down to around room temperature, where the solubility of phthalic acid is somewhere around one thirtieth that of the total solubility of phallic acid in boiling water.

So as my solution cooled slowly, the solubility of the phthalic acid drops dramatically. So, that particular compound now has nowhere else to go but out of solution. The key here is that I allowed the process to happen very slowly. So I guess it's a good thing I like to talk. Now, by cooling slowly, I've avoided trapping of impurity molecules in those growing crystals. And at the completion of my recrystallization, they'll have very large, well-formed, pure phthalic acid crystals that can simply be removed from solution by a technique like filtration.

So if I light my sample a little bit, you can probably see in the front and the bottom there I've got some very large, very well-formed crystals. And I'm going to try to retrieve those now, not using filtration, but instead, using a simpler technique to separate liquids from solids known as decanting. Now, decanting is actually quite simple. I'm just going to try to pour the liquid out from this beaker while leaving as much of the solid behind as possible. It just takes a steady hand and a little bit of patience to remove that liquid.

There we go. And as I begin to remove the liquid, it should become clear how pure these crystals really are. They're not only large and well formed, but save for just a little bit of the solution that they crystallized from, which is sometimes called the mother liquor, there's really not a lot of red to speak of. So, the impurity is now in the solution, which I have separated from the majority of my crystals, few of which I'm going to dig out now so that we can take a look at the difference in size and shape in comparison to what we started with.

So here are just a few of the crystals that formed. And when we compare them to not only the sample that we started with that was impure, we can clearly see that the color is a lot more white. But the crystals are even larger than the material that I started with before it was even fouled by this red compound. So, large crystals of the proper color tell me that I have at least, in part, successfully purified what was an otherwise impure sample. So washing these with a little bit of cold water will probably get a crystal that's almost as white and pure as the one that we started with before we ever added any dye at all. This is the power of recrystallization.

So, pure crystals tend to be more stable than their impure counterparts. Now, we've already looked at how this can be used to purify a compound by crystallizing it from a solution. But we can exploit this phenomenon in a completely different way by observing the melting behavior of a sample to help us identify it and also assess its purity. So let's take a look at that right now.

I'm going to bring back a crystal that we've looked at once already. That's the benzamide crystal. And again, I'm going to build this beautiful network of intermolecular attractive forces holding it all together, the hydrogen bonds, the van der Waals forces. Now, because there's so many forces holding it together, it's very difficult to take the crystal apart, not only by dissolution but also by melting, which involves heating. So, if I give these molecules a little bit of kinetic energy by adding some heat, let's take a look at what happens.

Well, they have a little bit of kinetic energy now. They want to be free to move. But they don't want to move badly enough, because there are too many forces holding them all together. But if I continue to add energy in the form of heat, I eventually reach a point at which there's enough kinetic energy contained in those molecules that they no longer can be held together by those forces within the crystal, and melting takes place. But if there's an impurity present, just as there was last time in our example, where we had benzamide with a little bit of benzaldehyde contained inside, well, remember, we had to sacrifice these specific intermolecular forces, meaning that overall there's less holding that crystal together. What this means from a bulk perspective is, as I heat up a crystal that has impurities and imperfections, like those missing intermolecular attractions, my weaker crystal should melt more easily. And indeed, in my example here, I add a little bit of heat, and that's all the kinetic energy it takes to get my molecules moving enough to overcome whatever forces are remaining holding it together.

So, just as dissolution involves liberating individual molecules from one another by solvation, so melting liberates them by heating them until they have sufficient kinetic energy to overcome those forces holding them together. With this in mind, the logic behind using melting to assess purity should become clear—that more purified crystals have greater forces holding them together, meaning they require more energy to melt. In other words, as purity increases, so does the melting point.

This concept is clearly illustrated using a hypothetical phase diagram for a binary system. Now, in order to construct this kind of phase diagram, we first take the phase diagram for a pure substance, which we could just call substance A. So here I've drawn one for you with all the phase transitions that we recognize from an introductory chemistry course, including boundaries for melting and freezing, boiling and condensing, and of course, sublimation and deposition. But this diagram only applies to a pure substance. So, to create a binary-phase diagram, we settle on a constant pressure, like one atmosphere, the pressure in most laboratories. And at that pressure, we project the diagram along a third axis of composition. When we do this for two hypothetical solids, which, in this example, we could call A and B, focusing on that melting transition, the result looks something like this.

So here's my phase diagram for a binary system at a given pressure. Notice that my horizontal axis is now mole percentage, meaning that on the left-hand side, I'm dealing with pure, in this case, green squares. And on the right-hand side, pure, whatever this molecule may be; we've called it red triangles for illustration purposes. But there's some very interesting features to this phase diagram that you don't see in the phase diagram for a pure substance. So, let's wipe this clean and rebuild the phase diagram piece by piece, investigating the effects of mixing one compound into another.

Let's start with this compound over here, my green squares, when it's absolutely pure. Now, I know that when it's absolutely pure, its melting point should be a single temperature. There is only one temperature at which a pure substance can coexist with itself in liquid and solid forms. So, if I were to heat a sample of this compound, I expect it to be solid up to its melting point, after which I expect it to transition into a liquid, and I can continue heating. But look what happens as I introduce a little bit of my other compound, in this case, as an impurity.

Well, two very interesting things happen. First is that the phase boundaries move downward. And the second is that there are actually two of them, not a single phase boundary anymore. So, what this means is that if I were to heat up a solid of a composition, let's say somewhere in this region where I have mostly my green squares but a little bit of my red triangles, take a look at what happens. I pass through another range of my diagram. It actually takes a range of temperatures to melt. Not only that, but both of them are lower than the original melting point of my green square compound.

We call this phenomenon melting point depression, and it occurs because one compound is acting as an impurity and the other disrupting the forces holding the crystal lattice together. And the broadening that takes place in the melting transition is known as incongruent melting, which is an effect of having one compound moving into the liquid phase faster than the other as melting takes place, thereby changing the composition of the solid as well. So there's a range of compositions which can coexist at liquid and solid forms for mixtures.

But then we reach an interesting point in the diagram; it comes back to a confluence at a point somewhere in the middle of the phase diagram, though not necessarily directly in the middle the way I've drawn it here. This is known as the eutectic point, eutectic meaning easy melting, essentially. And once we crossed that eutectic point, which, by the way, is the only composition at which a single melting temperature exists, we start to see the transition temperatures go back up until eventually they reach a confluence again at pure compound B or, in my case, red triangles.

So what's going on here? Well, what's going on here is we've reached a point where we can start to think of it as green squares are now the impurity in my red triangles. And so, my melting point goes back upward as my sample becomes purer and purer in the other compound. So I simply create a sort of a reflection. Although, the absolute temperatures and widths of these transitions are not necessarily the same on both sides, but these features are very consistent in well-behaved binary systems.

Finally, of course, we reach the point where we have our melting point for our pure, red compound, and again, we have a single-transition temperature. So, why do we need to understand that kind of complex phase behavior? Well, because it gives us a useful tool to confirm the identity and purity of a compound. We use a device commonly referred to as a Mel-Temp apparatus, which consists of a heating block with a temperature control and a viewing window.

Now, if I load a sample of my original crude, say, phthalic acid in one of my recrystallized materials, I can confirm that the melting point of the purified material should be higher. Not only that, but if I were to make a third sample consisting of my purified phthalic acid ground thoroughly with a small amount of a standard, that would allow me to confirm its identity, since adding anything but the same compound to itself should reduce the sample's melting point. If my sample were anything but phthalic acid, the melting point of the mixture with standardized phthalic acid should be depressed. We call this simple technique a mixed melting point analysis. And it's a very commonly used quick check for purity and identity when synthesizing compounds in a lab.

So, today, we started to think beyond simply making and characterizing a product to the question of how we might purify one. We discussed the effect of the presence of impurities on organic crystalline solids and how they tend to reduce the favorable intermolecular forces holding the crystal together, creating a crystal which is less stable. We took a look at how this observation can be used to devise a way to purify such materials by dissolving a sample to solve and separate all the constituent molecules, then cooling or concentrating the solution very slowly to promote formation of the most stable, pure crystal.

We discussed the process of melting an organic crystalline solid and how the inclusion of different molecules usually results in not only a depression of melting point, but also in a broadening of the melting transition resulting from a phenomenon called incongruent melting. Finally, we talked a bit about how we can use the phase behavior of such solids as a means of probing their purity and identity using a mixed melting point analysis.

Next time, we're going to turn our attention to another familiar phase transition—boiling. We'll explore how phase behavior of more volatile organics can also be used to purify and identify them. I'll see you for that discussion next time.

Purifying by Distillation
Lecture 33

D istillation has found many uses through the ages, from producing potable water and alcoholic beverages to refining oil and gasoline. It also finds use frequently in the organic chemistry laboratory as a method for isolating liquids of varying volatility from one another. In this lecture, you will learn about the fundamental laws governing this influential chemical technique, including Raoult's law, Dalton's law, and the ideal gas law. In addition, you will explore some advanced distillation techniques.

Vapor Pressure

- All liquids exist naturally in equilibrium with just a bit of their vapor, even at temperatures well below their boiling points. The exact amount of vapor varies from substance to substance and depends on how well molecules of the same kind stick together through intermolecular attractions.

- We measure the amount of vapor above a liquid in terms of the pressure that it exerts, so this property of liquids has come to be known as vapor pressure. The higher the vapor pressure of a given liquid, the faster and more easily it will convert to the gas phase. We often report vapor pressures in units of torr. One torr is equal to 1 mm of mercury in a barometer.

- Increasing the temperature of a sample means increasing the kinetic energy in the molecules making it up. So, increasing the temperature of a liquid sample gives more of its molecules enough kinetic energy to escape the liquid and become a gas. This increases the vapor pressure of the sample.

- When a liquid is heated to a temperature at which its vapor pressure is equal to the externally applied pressure, such as the 760 torr of pressure exerted by the Earth's atmosphere, the liquid boils. When the pressure we are working against is 1 atmosphere, we refer to this temperature as the normal boiling point.

Raoult's Law, Dalton's Law, and the Ideal Gas Law

- Raoult's law, named for French chemist François-Marie Raoult, states that the vapor pressure exerted by a component in a mixture of miscible liquids is equal to its vapor pressure when pure multiplied by its mole fraction in the mixture. In other words, the vapor pressure of each component is proportional to the fraction of sample molecules it makes up.

- Dalton's law, named for John Dalton, tells us that the sum of the partial vapor pressures exerted by all of the components of such a system is equal to the total vapor pressure exerted by that system. In other words, vapor pressures are additive. So, varying the amount of each liquid making up the mixture can alter its vapor pressure.

- The ideal gas law is actually a combination of several laws, relating pressure, volume, abundance, and temperature through a constant called the ideal gas constant. The equation for the ideal gas law is $PV = nRT$, where P is pressure, V is the volume occupied by the gas, n is the number of moles of gas, T is the temperature of the vapor, and R is the ideal gas constant. Because R is constant for all systems, we can assert that PV/nT for any gas or collection of gasses should always be equal.

- The ratio of the partial pressure of a gas to the total pressure must be equal to its mole fraction in the sample. The vapor above miscible mixtures of liquids is more enriched in the more volatile component or components.

The Distillation Experiment

- If we could boil the mixture and channel the vapor somewhere else, condensing it in a separate container, we would have a liquid of greater purity than we had started with. It is this process of boiling, transporting, and then condensing a sample that adds up to create the technique of distillation.

- However, this process creates a distillate that is enriched but not purified enough for certain applications. So, how would you get a distillate of higher purity?

- Usually, the first impulse is to simply redistill the distillate, thereby obtaining an even more enriched, higher-purity distillate. The problem with this approach is that it is very labor- and energy-intensive. It requires either a major modification to the still, adding a second condenser and receiving flask, or that we stop, clean the entire apparatus, recharge the boiling flask with the distillate, and run again.

- Neither of these options is particularly attractive, especially when there is an easier way—a technique known as fractional distillation. A fractional still differs from a simple still in just one crucial way: It contains a vertical column between the boiling flask and the still head. It is often packed loosely with inert material like glass or metal, which increases the surface area inside the vertical column.

- The ultimate effect of all this is that a single distillation runs with greater efficiency than a simple still can achieve. If the distillate is as pure as that obtained by two simple distillations, we say that the still runs at two theoretical plates. This is a reference to early fractionating column designs, which actually contained plates along the length of the column, on which one cycle of condensation and vaporization could take place before the vapor moved on to the next-highest plate.

Azeotropes

- As powerful as distillation is as a technique, it does suffer from limitations. For example, grain alcohol is 190 proof—meaning that the mixture is 95% ethanol and 5% water by volume—not 200 proof, or 100% alcohol for a very good reason: Ethanol and water cannot be distilled beyond this proportion.

- Raoult's law, Dalton's law, and the ideal gas law are predictions based on ideal behavior, in which gas molecules do not interact with one another significantly. The truth is often quite different from this. Because these laws are not rigorously accurate models, we often see distortions in the behavior of systems that we are trying to distill.

- The mixture of ethanol and water is actually a rather mild example of this. The liquid-vapor phase diagram for this system has a small local minimum at about 95% ethanol. It looks a lot like the eutectic point, which is the only composition at which the liquid and solid are of the same composition.

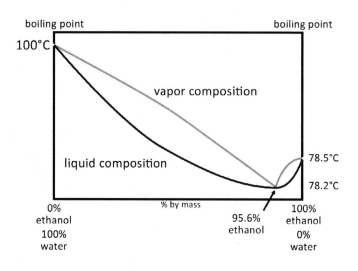

- When the analogous feature is seen in a liquid-vapor system, we call the mixture an azeotrope, derived from Greek terms meaning "no change on boiling." It is a very appropriate name, because the vapor above the liquid is of exactly the same composition as that liquid, making distillation of such a mixture futile.

- A fractionating column that reaches the Moon would not be sufficient to obtain pure ethanol by distillation. We would have to resort to other means to chemically dry the azeotrope and make it 100% ethanol. From the perspective of an alcoholic beverage company, it simply isn't worth all of the extra effort to remove that last little bit of water from a product that is most likely going to be mixed right back in with another aqueous solution to prepare a drink.

Advanced Distillations

- Since a century or two ago, many additional methods for distillation have been developed to create new spins based on this old technique. While the simple and fractional distillation techniques are designed to carefully coax one volatile organic away from another, there are situations in which we can be a bit more heavy-handed. An example is the rotary evaporator.

- For example, let's say that we want to recover a nonvolatile solute from solution, such as nonvolatile dye molecules dissolved in a sample of acetone. Essentially, one of our two components has no vapor pressure whatsoever. In this situation, we want to use a system that can quickly and quantitatively distill away all of the volatile material, leaving behind only the nonvolatile dye molecules.

- Another potential sticking point when designing a distillation is how to distill a material that is known to decompose before it reaches its normal boiling point. This can be accomplished in a number of ways, but when the desired compound is immiscible with water, chemists often turn to the technique of steam distillation to get the job done.

- Take the example of eugenol, which is the primary component of the essential oil of cloves and is prized for its anesthetic properties and as a flavoring agent. But it has a very low vapor pressure, giving it a normal boiling point of about 254° Celsius. Such a high boiling point makes eugenol a poor candidate for purification by distillation.

- However, there is one critical property of eugenol that we have not yet considered. It is immiscible with water. This means that Raoult's law does not apply to a mixture of eugenol and water. Instead, water and eugenol will each establish their vapor pressure independently of their mole fraction in a mixture.

- At 100° Celsius, eugenol has a vapor pressure of about 4 torr. This means that in the presence of boiling water, eugenol will only make up about 0.5 mole percent of the molecules in the vapor phase.

- This might seem like a hopeless endeavor, creating a vapor that is only 1 part eugenol and 199 parts water. But keep in mind that this approximation is a mole percentage and not a mass percentage. We can easily convert it into mass percentage, multiplying by the molar mass of eugenol and dividing by the molar mass of water.

- This leads us to the conclusion that the vapor above a mixture of boiling eugenol and water will actually have 4 mass percent eugenol—still a small amount, but not nearly as disappointing as our mole percentage made it seem.

- So, by boiling whole cloves in water, we can generate a vapor that is 4% eugenol in water at a very safe 100° Celsius. Collecting this vapor using a West condenser gives us a mixture of eugenol and water, which are easily separated because they are not miscible.

- What makes steam distillation particularly useful in situations like this is that the steam need not be exactly 100° Celsius. More complex systems can be set up to introduce steam of varying temperatures, allowing us to maximize the yield of oils without losing product to degradation. This technique is very commonly used to extract oils like that of the clove plant because they tend to be low-volatility, water-immiscible materials.

Suggested Reading

French, *The Art of Distillation*.

Saltzman, M. D., "The Art of Distillation and the Dawn of the Hydrocarbon Society," *Bulletin for the History of Chemistry* 24 (1999): 53–60.

Williamson, *Organic Experiments*, Chap. 5.

Questions to Consider

1. In what situations is simple distillation most likely to be sufficient to separate two miscible liquids from one another?

2. How does packing a fractionating column with solid material serve to improve the efficiency of fractional distillations?

Purifying by Distillation
Lecture 33—Transcript

Potent alcoholic beverages from the waste product of yeast, drinkable water from the sea, gasoline from crude oil, humanity has always had a deep interest in the purification of liquids to obtain commodities of great value. One of the earliest and most effective techniques to purify liquids relies on two simple physical changes, boiling and condensation, which combine to create a familiar technique called distillation.

Most of us are predominantly familiar with this technique as a method of producing beverage alcohol. This application has been employed for so long that its true origins in human history are lost to the ages. What is certain is that many early civilizations independently discovered this trick, which can be used to produce a beverage from such varied sources as fermented rice, bananas, or even milk.

But ethanol is just one of many liquids which we find particularly useful in its purified state. In the 4th century B.C., Aristotle observed this phenomenon related to water and wrote on it in his work, *Meteorologica*. He wrote, "Salt water, when it turns to vapor, becomes sweet. And the vapor does not form salt water again when it condenses. Wine and all fluids that evaporate and condense into a liquid state become water." Now, by "sweet," what Aristotle meant was drinkable. He had observed that sea water, wine, and all other common solutions of his day could be made into fresh potable water simply by boiling it, then re-condensing the vapor in a new container. Now, clearly seawater and fresh water were made from the same basic substance and modified only by some admixture, which could be removed by this process.

Of course, in his day, Aristotle only had access to aqueous solutions, so his observation that all fluids become water in this process would be a bit erroneous from a modern perspective. But he did demonstrate an understanding that liquids can be vaporized, collected, re-condensed to form a liquid of a different composition than the original, a primary tenant of the technique of distillation.

Aristotle's observations weren't merely academic either; his ideas were applied in the centuries following his death. For example, Greek sailors, according to the accounts from Alexander of Aphrodisias, boiled seawater and suspended large sponges from the mouth of bronze vessels above them to obtain drinking water. Now, here Alexander's describing the desalination of seawater.

In the middle of the 1600s, an Englishman and physician named John French played a critical role in thrusting distillation from the realm of an alchemical art to a chemical science. He very eloquently writes in his now-famous work, *The Art of Distillation*, "I shall not stand here to show where the art of distillation had its origin as being a thing ... little conducing to our ensuing discourse. But let us understand what distillation is." In yet another of my favorite quotes of all time, French writes, "Before you take yourself to the work, propound to yourself what you seek and enter not upon the practice until you are first well versed in the theory." A more telling quotation of those times could not possibly exist.

French has always been on my list of top-10 scientists I'd like to sit down for a beer with, though perhaps, in his case, a whiskey would be a more appropriate drink. In his time, French's work was the most influential universal compendium of distillation theory and techniques that was ever published, a true asset to the growing science of chemistry. Today, it offers us a fascinating insight into the change in thinking which was going on during his day, a shift from the artistic and romantic descriptions of the alchemical world to the factual and detailed observations of the chemical world.

But distillation is not simply an ancient technique relegated to the pages of history. Modern fuels are a particularly pertinent example of a more recent societal change made possible by this purification method. A fantastic example of this is the refining of crude oil into useful materials. And we owe a great deal of that practice to J.D. Rockefeller's Standard Oil Company and his competitors. You see, there's a good reason why we refer to fossil fuels as petroleum distillates. It's because the refining process primarily relies on distillation to separate components of crude petroleum, based on little more than their boiling points. Rockefeller's true genius was not to employ the technique; that had been done long ago. But what really helped to make

Rockefeller successful was his realization that, in addition to lamp oil, he could use distillation to collect and sell many of the other products of various volatilities which other companies simply treated as waste, and this included octane for gasoline. So just imagine that; gasoline was once considered a waste product.

So distillation has found many uses through the ages, from producing potable water and alcoholic beverages to refining oil and gasoline. It also finds use frequently in the organic chemistry laboratory, as a method for isolating liquids of varying volatility from one another. The applications of distillation go on and on, but in this lecture, we'll take French's advice and first take a look at the fundamental laws governing this influential chemical technique.

First, a bit of background on the process of vaporization. All liquids exist naturally in equilibrium with just a little bit of their vapor, even at temperatures well below their boiling points. Now, the exact amount of vapor varies from one substance to the next and depends on how well molecules of the same kind stick together through intermolecular attractions. For example, under a given set of conditions, water will exert a lower vapor pressure than will acetone, because the water molecules can hydrogen bond to themselves, causing them to prefer the liquid phase more than the gas. This is easily observed; just place a drop of water next to a drop of nail polish remover on a countertop and see which one evaporates first.

We measure the amount of vapor above a liquid in terms of the pressure that that vapor exerts. So this property of liquids has come to be known as vapor pressure. The higher the vapor pressure of a given liquid, the faster and more easily it will be converted to the gas phase. Now, we often report vapor pressures in units of torr. One torr is equal to one millimeter of mercury in a barometer. So 760 torr is equal to one atmosphere of pressure, the likes of which we would experience on a lab bench.

Now, increasing the temperature of a sample means increasing the kinetic energy in the molecules which make it up. So increasing the temperature of a liquid gives more of its molecules enough kinetic energy to escape the liquid and become a gas. This increases the vapor pressure of the sample. When a liquid is heated to a temperature at which its vapor pressure is equal to the externally applied pressure, like the 760 torr of pressure exerted by the Earth's atmosphere, for example, the liquid boils. When the pressure we are working against is one atmosphere, we refer to this temperature as the liquid's normal boiling point. This is most easily observed by heating a sample of pure water. At room temperature, water's vapor pressure is only about 3% that of atmospheric pressure. But as the temperature increases, more and more water vapor can be seen forming at the liquid's surface until the water reaches 100°C, at which point its vapor pressure is now one atmosphere, or 760 torr, and boiling takes place.

But that's pure water. And our goal today is to show how we can separate mixtures of liquids from one another using this process. So let's call on a few more great minds from the 1800s to help us with that. Raoult's law, named for French chemist Francois-Marie Raoult, states that the vapor pressure exerted by a component in a mixture of miscible liquids is equal to its normal vapor pressure when pure, times its mole fraction in the mixture. In other words, the vapor pressure of each component is proportional to the fraction of sample molecules it makes up. So a solution which is 50% water at room temperature will only exert a vapor pressure of 1.5 torr, instead of three torr.

Dalton's law, named for John Dalton, tells us that the sum of the partial vapor pressures exerted by all of the components of such a system is equal to the total vapor pressure exerted by that system. In other words, vapor pressures are additive. For example, if our water was diluted to 50% using ethanol, we would have to include the vapor pressure of the ethanol in the total vapor pressure of our mixture as well. Now, ethanol's vapor pressure is about 50 torr at room temperature, so it contributes 25 torr to the vapor pressure of the mixture, according to Raoult's law, giving the mixture about 26.5 torr total vapor pressure, in accordance with Dalton's law. So varying the amount of each liquid making up the mixture can alter that mixture's vapor pressure.

Finally, we're going to use ideal gas law, which is actually a combination of several other laws relating pressure, volume, abundance, and temperature, all through a constant called the ideal gas constant, which we represent by the letter R. So let's start our discussion about the process of vaporization and how it applies to distillation in an object lesson using benzene and toluene, two fairly simple hydrocarbons. Now, I'll be using numbers for toluene and benzene, but I'll be representing the molecules as blue spheres and red spheres throughout this demonstration so that we keep our attention on the physical behavior of the molecules and not the structures of them themselves.

So here I have a beaker that's filled with benzene and toluene. And this mixture of benzene and toluene is 50 mole percent of each. Now, that means that if I model my blue spheres as toluene and my red spheres as benzene, I can begin to follow them as they change from the liquid to the vapor state. Now, at a given temperature, the vapor pressure of toluene is normally 300 torr, and at the same temperature, the vapor pressure of pure benzene is normally 1,200 torr. So I can use Raoult's law to calculate the vapor pressure exerted by each in my mixture. For example, the vapor pressure exerted by toluene is equal to Chi, the Greek letter Chi, which is its mole fraction, times the vapor pressure it would exert if it were pure. In other words, the toluene should exert a vapor pressure of 150 torr. Now, a similar calculation for benzene allows me to predict that its vapor pressure above the liquid should be equal to 600 torr. And a quick check using Dalton's law tells me that this liquid will be boiling, maybe on a cloudy day here in Washington, DC, when the atmospheric pressure is 750 torr, which equals the vapor pressure.

Now here's the interesting part. If I take the vapor pressures of my liquids and use the ideal gas law, which allows me to predict that the pressures they exert is proportional to the number of moles in the gas phase, I calculate that the mole fraction of benzene in the vapor phase is 80 mole percent, not 50 mole percent. So this means that, in the head space above my liquid, whatever vapor does form at this temperature is actually going to be enriched in benzene, which is the more volatile of the two components. Now, this is very important to distillation and how it allows us to separate liquids.

So, let's say that I could take a sample of benzene and toluene and place it in an apparatus in which it can vaporize and in which that vapor is then channeled into a new region in space where it is then condensed back into a liquid. Well, if I can do that, then the liquid I collect should be of the same composition as the vapor in that original sample. So we use a device like this called a simple still, which consists of what we call a boiling flask, which is heated, a still head, whose job is to direct the vapor into another downward facing region, which we call a West Condenser.

Now, the West Condenser has a water jacket to keep it cold, which induces condensation. But that condensed vapor rolls downhill into a new receiving flask, and it's directed that way through a flow controller or a vacuum adapter in this case. So knowing what we know about how a mixture of benzene and toluene will vaporize, if I run my still by adding heat to the boiling flask, I expect that the vapor which is moving through the system will be enriched in the benzene, and therefore, at the end of my distillation run, when I remove my receiving flask, it should contain 80 mole percent benzene. So I've been enriched my sample in the more volatile component through distillation.

So now that we've talked a little bit about the theory behind simple distillation, let's take a look at one in progress. So I've constructed here a simple distillation apparatus, the type of which you might find running in an organic chemistry lab. Currently, I'm distilling a mixture of 50-50 ethanol in water. Now, the apparatus consists of not only the elements we looked at previously in our slides, but a few other practical pieces that help it work.

The first is this device called a lab jack, and the lab jack gives me a little bit of clearance so that I can raise and lower my heat source when I need to. On top of the lab jack is a stir plate. Now, this is also a heating plate, but I'm not using the heat, because remember, we talked about this in our previous lecture, that it's not a good idea to put anything with organic solvents directly on top of a hot plate. And since I'm working with ethanol, which, in certain concentrations is flammable, I can't do that. Instead I've used a device which is called a thermal well. A thermal well is a ceramic heater, and I'm controlling it with a little controller in the back here, which is hidden from view, which regulates how much current is reaching the thermal well. So that's how I set the temperature of the still just the way I like it. Now, right now, I have a round-bottom boiling flask, which is clamped in place. So this is the lowest piece of glassware in my apparatus, and so I've clamped it on a ring stand so that, when I lower the lab jack, my heat source is removed from my organic sample. So this is all about safety up to this point.

Now, about the distillation. The boiling flask contains a mixture of boiling water and ethanol. And if you look very closely, you can actually see the vapor condensing within the still head here. And a great deal of it is falling back into the boiling flask, and that's OK. But some of that vapor traverses the still head into this arm, which is in a downward-sloping direction, and is in line with a water-cooled condenser. So this condenser here, actually, if I wipe away some of the condensation that's on it, you might be able to see this, has two concentric tubes. The outside tube is actually plumbed to cold water, which is circulating because of a pump that I have underneath the table. So it's constantly keeping the inner portion of this column cold. That's why the vapor is condensing in this cold region and rolling downward where my vacuum adapter is directing the flow into a receiving flask. You may notice that it's dripping downward, and that's actually the origin of the term "distill." It's from the Latin meaning to drop downward.

So if I'm really enriching my sample here in ethanol, I should be able to detect that in the temperature of the vapor as it moves through the still head. So I have an infrared thermometer here, and I have placed a few white stickers on my apparatus so that I can get the proper reflection of the IR light to measure the temperatures going on in this still. Now, if I point my thermometer directly at the boiling flask, I see that at least the surface of the boiling flask is somewhere around 80°C, 77 according to my thermometer. But if the vapor, which is moving into the condenser, is of a different composition, then it should be a different temperature. If I've increased the concentration of the more volatile component, I should see a decrease in temperature as it moves up into the still head. So let's take a look at the still head temperature, at least on the outside of the still head. See it's moving around quite a bit as the vapor sort of moves and swirls inside, but my reading never exceeds about 70°.

So, even though I've been measuring the temperatures of the outside of the still, or the glass surface, rather than the actual liquid and vapor, we can see the trend occurring here, that it's much warmer down here where the boiling liquid is compared to the temperature of the vapor, which is moving through the still head. Now, if this were a pure liquid, that should not be the case, because boiling means that the liquid and the vapor are in thermal equilibrium with one another, which clearly they aren't. And that is the source of the purification that we get during a distillation.

So we've seen how we can separate, or at least enrich, a sample in a more volatile component using simple distillation. But what if we want a really highly purified sample of a particular liquid? Remember, in our calculations before with benzene and toluene, we determined that we would only get a distillate of about 80 mole percent benzene. But what if we need one that's 90, or 95, or even practically 100 percent benzene? Well, we have a tool we can use to predict exactly how liquids and vapors will behave to determine the outcome of a distillation and whether or not we need to take extra measures. Now, for that, we use something called a liquid-vapor composition plot. And a liquid-vapor composition plot, simply put, is a phase diagram for a binary liquid system at boiling, as opposed to a binary solid system, which is melting like we looked at last time.

So, remember we were operating at a specific temperature, boiling a mixture of benzene and toluene, which was 50 mole percent benzene, but created a vapor that was 80 mole percent benzene. So, at that given temperature, on my plot, I can place those two percentages of benzene, here 50 for the liquid, and here 80 for my vapor composition.

But these calculations and this experiment have been done many, many times over at many, many different compositions of benzene and toluene. So if we plot similar results for different compositions at their boiling points, we start to see a pattern emerge within the plot. And connecting the dots gives us what looks a lot like our phase diagrams for a binary system that we've seen before. So this is the origin of the liquid-vapor composition plot. It's a tool which allows me to predict the outcome of a distillation without ever having to go into the lab and run one.

So let's say that I really do want a very, very pure sample of benzene. Can I use distillation to achieve that? Well, our calculations would suggest otherwise for a 50 mole percent mixture with toluene. But what if I were to collect that distillate and then re-distill it? Well, I could build an apparatus like the one I've shown you here in which I have a boiling flask and then an intermediate class which is a collection flask for a first round of distillation but is a boiling flask for a second round of distillation.

Now, if I were to run a system like this, I would expect my 50 mole percent benzene to collect in the first flask as an 80 mole percent benzene solution. But that 80 mole percent benzene solution, when heated, I would expect to distill yet again and form a new condensate which is even more enriched. Now, just how enriched it is I can determine using my liquid-vapor composition plot. Now, it turns out, it's 97 mole percent. So let's see how I figured that out.

Using the liquid-vapor composition plot, I started with a liquid of 50 mole percent, which gives me a vapor of 80 mole percent, which gives me a condensate of 80 mole percent in the receiving flask. Now I'm going to heat that 80% liquid that I have re-condensed. That creates a vapor which re-condenses to form a liquid with a composition of 97 mole percent benzene. Now, I've set it up here as two sequential distillations. But we'd really like to be able to do this in a system in which we can run a single run without having to build all of this, or take it apart and clean all of this at the end, or use two different sources of heat. So, in short, is there a simpler way that we can do this?

The answer is yes, and we call it a fractional distillation. This is a fractional distillation scheme, and I've really only changed my simple still by one very important thing; I've added what's called a fractionating column to it. And the insertion of this fractionating column gives lots and lots of surface area onto which my vapor, as it ascends the column, can condense and then re-vaporize and then condense and re-vaporize as it climbs up the column. So if I charge this flask with 50 mole percent benzene and then run my fractional still, let's take a look at what goes on inside this region, this fractionating column that I've added.

As I heat the sample, I'll see vapor beginning to ascend, but every time it condenses along the walls, falls into a slightly warmer region and re-vaporizes, it becomes more and more enriched in the more volatile component, in our case, benzene. Near the base of that column, it's as though I were doing a simple distillation; I get about 80 mole percent. But if you look carefully about midway up that column, you'll notice that far fewer of my toluene molecules are making it to this location on the column. And if you go even higher, near the very, very top, really none of them are. So if I run this particular column as I've depicted it here, collecting the vapor that exits through the top of the column, I expect to get what is essentially 100% pure benzene in a single distillation, starting with 50 mole percent benzene.

So here's how I was able to do that. I started with 50 mole percent, but my column here is running not as though a single round of distillation had taken place, creating an 80% benzene distillate, but rather, as though a second one had taken place, giving 97%, and then yet again, a third one had taken place, finally getting me right around the limit of my diagram here, to about 100% benzene.

Now, when I do a single run in a fractional still, which gives me a distillate of a purity that resembles that of something that's been distilled twice or three times, I report on the efficiency of my still using a term called theoretical plates. Simply put, a 97% pure distillate would mean that my fractional still operates at two theoretical plates, because it's as though I had distilled and re-distilled a second time. If, instead, my fractional still gives me 100% pure benzene in the first run, then it's three theoretical plates, because it's as though I've gone over and down three times on my liquid vapor composition plot.

So I've made a small change to my simple still setup from last time, inserting not only the fractionating column itself, which is here, but I've also, you'll notice, packed a little bit of steel wool inside. And the purpose of that packing is to give extra surface area onto which multiple rounds of condensation and vaporization can take place as my sample ascends the column. In fact, if you look very closely, you'll see plenty of condensation falling back down, because it's formed before the vapor actually reaches the side arm. So only the most volatile of components can get to the top.

Now, again, in this case, I have 50-50 ethanol and water in my boiling flask. So if I take a look at the temperature of the boiling liquid, it's no surprise that the temperature is somewhere around the mid 80°C. But, as my sample ascends the column, let's say about halfway up, I've got another sticker here so I can measure my temperature. Now, in this case, the temperature of the ascending vapor is no longer in the high 80s; now it's about 80°. And finally, at the top, if I measure the actual temperature of the vapor which is reaching my West Condenser, I see that that vapor has a temperature of about 75°.

So what I've shown you here is proof that, as the vapor ascends, the boiling point of the associated combination of ethanol and water continually decreases. Now, knowing that the boiling point of ethanol is lower than that of water, it should come as no surprise that my distillate, that is, that liquid which is dripping down from the far side, is dramatically enriched in ethanol. In fact, it's just about 95% ethanol by volume.

Powerful though distillation is as a technique, it does suffer from limitations. You may have seen this product for sale before. It's a grain alcohol beverage which is labeled 190 proof. Now, that means the mixture is 95% ethanol by volume and 5% water. If you ever saw this product on the store shelf, you might have wondered to yourself, if you're going to make alcohol that strong, why not just go for broke and make it 200 proof or 100% absolutely pure alcohol? Now, there is, in fact, a very good reason that the manufacturer decided to stop at 95%. It's because ethanol and water simply cannot be distilled beyond that proportion. You see, Raoult's, Dalton's, and ideal gas laws are just that, predictions based on ideal behavior in which gas molecules don't interact with one another significantly. And, as you can imagine, the truth is often quite different from that.

So because these laws are not rigorously accurate models, we often see distortions in the behavior of systems which we're trying to distill. Ethanol and water is actually a rather mild example of this. The liquid-vapor phase diagram for this system looks like this. With a small local minimum at about 95% ethanol, it looks an awful lot like something we saw earlier, doesn't it? Right. It looks just like the eutectic point in melting of solids, which was very interesting, because it was the only composition in which the liquid and solid were of the same composition.

When the analogous feature is seen in a liquid-vapor system, we call that mixture an azeotrope. This is derived from Greek terms, meaning no change on boiling. It's a very appropriate name, because the vapor above the liquid is of exactly the same composition as that liquid, making distillation of such a mixture futile. Now, a fractionating column which reaches the moon would not be sufficient to obtain pure ethanol by distillation. We have to resort to other means to chemically drive the azeotrope to make it 100% ethanol. Now, from the perspective of an alcoholic beverage, it simply isn't worth all of the extra effort to remove that last little bit of water from a product which is most likely going to be mixed right back in with another aqueous solution to prepare a drink.

So let's sum up what we've covered today. We started with a mathematical explanation for why vapors above boiling mixtures tend to be enriched in the more volatile component. We relied on the concept of vapor pressure and Raoult's, Dalton's, and ideal gas laws to support our prediction. We used the results of these calculations to construct a phase diagram, which is commonly called a liquid-vapor composition plot, and we explored how these plots can be used to graphically predict the outcome of a distillation.

Next, we investigated a commonly used simple distillation apparatus, which diverts vapor from boiling mixtures into a condenser, which then directs the re-condensed liquid into a separate collecting flask. We also looked at a fractional distillation apparatus, which differs by the insertion of a special column between the boiling flask and still head, allowing multiple rounds of vaporization and condensation to take place in a single run. This increased efficiency can be compared to the liquid-vapor composition plot to give it a numerical value in the form of theoretical plates.

Next, we covered the phenomenon of azeotropes, which are undistillable mixtures of liquids which form because of non-ideal behaviors. We saw how they manifest themselves as a local minimum within the liquid-vapor composition plot, making the vapor and boiling liquid identical in composition, thereby rendering distillation ineffective.

So over our last two lectures, we've covered all of the phases which we commonly encounter in the lab—solid, liquid, and vapor—and how transitioning from one state to the other can be used to purify and sometimes even identify substances in the lab.

In our next lecture, we'll return our attention to a different physical property, solubility. We'll see how this property can also be used as a very effective tool for isolation of non-volatile organic compounds in a process known as liquid-liquid extraction. I'll see you then.

Purifying by Extraction
Lecture 34

When a compound does not have the necessary physical properties for recrystallization, or if it is not the dominant material in a mixture, we can't turn to recrystallization to obtain an effective separation. But there are other ways to isolate organic compounds from one another based on a phenomenon known as partitioning. In this lecture, you will learn about one of the most common classes of techniques based on this principle: extraction. Specifically, you will learn how solubility can be used as a very effective tool for isolation of nonvolatile organic compounds in a process known as liquid-liquid extraction.

Partitioning in the Laboratory

- The contents of a bottle of salad dressing include water and olive oil. We can see these two ingredients because they create two distinct phases. Because the oil and water are immiscible, one simply floats on the other. In this case, it is the oil above the water. Also in the bottle are salt and some herbs. To simplify, the recipe contains just those four components: water and oil as solvents and salt and oregano oil as solutes.

- The water and oil provide two different chemical environments in which the sodium and chloride ions from the salt and the organic carvacrol from the oregano can dissolve. Knowing what we know about intermolecular forces and the role they play in solubility, we expect to find the ions of the salt preferentially dissolved in the high-polarity water, which can orient its

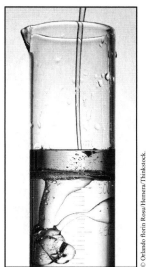

Oil and water create two distinct phases.

bond dipoles to create favorable interactions with the dissolved ions. The low-polarity carvacrol from the herb oils dissolves in the low-polarity oil layer on top, because the larger oil molecules are more polarizable and can form dispersion forces that solvate the carvacrol.

- It is exactly because of this separation that the proper blend of flavors is only obtained when we vigorously shake the bottle just before use. To pour just the top layer would be to put only the herb oils onto the salad, while to pour just the lower layer would be to only add the salt and acid flavors. When we do not shake the dressing—but, rather, deliberately allow the two phases to stay separate, dispensing each individually—we are doing a very simple form of liquid-liquid extraction.

- In this system, we assumed that all of a particular solute accumulated in one of the two phases available to it. We were able to make this assumption because of how drastically different sodium chloride and carvacrol are, being ionic and organic, respectively.

- But the truth for most organic molecules in a system like this is not so simple. Many of them have an appreciable solubility in both water and a certain organic solvent. When this is the case, we can expect a solute to establish a set of equilibrium concentrations in both solvents based on its solubility in each.

- This effect is called partitioning, and the equilibrium constant that governs it is called a partitioning coefficient. Partitioning coefficients are easily calculated from the maximum solubility of the solute in each of the two solvents. They give a numerical value that allows us to predict how much of each solute will be in each layer at equilibrium.

Effect of pH on Partitioning

- Compounds of varying solubilities can be concentrated or reduced in samples with a well thought-out liquid-liquid extraction. But there is a very powerful trick that we have not yet considered—that many organic compounds are titratable, meaning that they can be ionized by protonation or deprotonation in suitably strong acid or base, respectively.

- Partitioning depends on relative solubility of the solute in each solvent, so if we can change the solubility of the compound in one layer, we can change the partitioning coefficient. The key to understanding acid–base extraction is to recognize that when ionized, organic compounds tend to strongly favor dissolution in aqueous media because of its high polarity.

- Let's think about a simple organic acid with the formula HA. Its acid dissociation reaction is governed by the equilibrium constant K_a for this compound. An expression that relates the ionization state of the compound to the ratio between the dissociation constant K_a and the proton concentration in the water is $K_a = [H^+][A^-]/[HA]$.

- The Henderson-Hasselbach equation relates the ratio of acid and conjugate base through the difference in pK_a and pH: $pH = pK_a + \log_{10}([A^-]/[HA])$. Using this relationship, we can generate a plot of the percent ionized in water as a function of pH.

- The logarithmic relationship leads to a situation in which the acid is 99% in its neutral protonated form when the pH is two units lower than its pK_a. Conversely, it is 99% in its charged conjugate base form when the pH is two units higher than its pK_a.

- This means that we can maximize its partitioning into an organic phase by lowering the pH of the aqueous phase using a strong mineral acid, or we can maximize its partitioning into the aqueous phase by raising its pH using a base.

- For example, let's say we have phenol, a weak organic acid with a pK_a of about 10 that can deprotonate to become phenolate. If we were to mix a solution of phenol in ether with an aqueous solution at pH of 7, we create a situation in which the phenol can be neutral in the organic layer or neutral in the aqueous layer. It will prefer the organic layer.

- But if instead we use an aqueous solution at pH 12, the phenol has a choice between being neutral in the organic layer or a being a charged phenolate in the aqueous layer. This time, the aqueous layer competes much better for the solute, and the partitioning behavior of that solute changes.

Acid–Base Extractions

- The effect of altering the pH in an extraction can be significant and is fairly predictable, allowing us to tune an extraction to our liking in many cases. Let's again take the example of phenol, but this time let's assume that we need to separate a sample of phenol from a physical mixture containing benzoic acid. Both are similar aromatic compounds with similar polarities.

- If we were to dissolve the mixture in ether and then attempt an extraction with an aqueous solution at pH 2, because we are below the pK_a of both compounds, we expect them each to preferentially accumulate in the ether layer. The separation is not optimized.

- But say that we instead buffer the aqueous layer to a pH of 7. This time, we are three units above the pK_a of benzoic acid but three units below that of phenol. We are in a range of the Henderson-Hasselbach plot, which shows that benzoic acid will dissolve in the aqueous as benzoate, but the phenol will have to be neutral. So, the benzoic acid is expected to partition into the aqueous layer to a much greater extent.

- Finally, let's consider what would happen if the aqueous layer were pH 12. We are now well above the pK_a of both compounds, meaning that not only can benzoic acid dissolve in the aqueous layer as its charged conjugate base, but so can phenol. Both compounds prefer the aqueous layer, and the separation will again suffer.

- So, the selection of an aqueous layer pH must be carefully considered when attempting to isolate titratable organic compounds from one another.

- Yet another consideration comes up when dealing with organic bases, such as aniline. Aniline is an aromatic amine, which is in fact basic. As a base, its conjugate acid is anilinium ion, which has a pK_a of about 5.

- But because aniline is a base, its conjugate acid will be charged, meaning that its Henderson-Hasselbach plot will be inverted on the y-axis, reflecting its high water solubility in more acidic solutions. If we were to instead need to separate this base from phenol, the effect of this on pH selection is dramatic.

- We now must use an aqueous layer with a pH less than 3 or higher than 12 to reach a position in which one of the two components ionizes well in water. So, the drawback to separating acids from bases is that we must use extreme pH conditions, but the benefit is that we now get to choose which compound accumulates in the aqueous layer and which accumulates in the organic layer.

Extraction from Solids
- Partitioning can take place between any two distinct phases, not just two immiscible liquid phases. Some other examples are solid-liquid or liquid-gas systems. So, naturally, it should be possible to conduct extractions using such systems as well.

- Probably the most obvious example of an extraction using solid-liquid partitioning is the process of brewing a cup of tea. As we add hot water to a tea leaf, certain compounds dissolve into the water better than others. The large and insoluble material making up the tea leaves can interact with the caffeine, polyphenols, and other compounds that would otherwise be soluble in cold water.

© VvoeVale/iStock/Thinkstock.

We can manipulate the partitioning coefficients of compounds in solid-liquid systems using temperature; tea connoisseurs will understand this well.

- However, steeping tea in cold water leads to a very weak solution. So, just as we can manipulate the partitioning coefficient of a liquid-liquid extraction with pH, we can manipulate the partitioning coefficient of a compound in a solid-liquid system using temperature.

- Green tea is famous for its delicate flavors and aromas, but it also contains particularly high levels of tannins, a polyphenolic compound with a bitter taste and dry mouthfeel. So, the goal when preparing this delightful beverage is temperature control.

- Fine green teas require careful attention to brewing temperature, because water just below boiling partitions the pleasant and aroma-giving compounds into the water effectively—but water too close to boiling will separate the polyphenolic compounds that have a more bitter taste, ruining the otherwise enjoyable experience of a well-crafted green tea.

Williamson, *Organic Experiments*, Chap. 8.

1. For a given amount of extraction solvent, is it better to conduct one large extraction or several smaller extractions, pooling the extract at the end?

2. How does one determine the best-possible aqueous layer pH to separate two compounds by liquid-liquid extraction?

Purifying by Extraction
Lecture 34—Transcript

Previously, we discussed the method of recrystallization as a means of purifying non-volatile, organic, crystalline solids. Now, recrystallization is a powerful technique for the isolation of products when they have melting points well in excess of a solvent's boiling points, and of course, when the desired compound is present in great abundance, ensuring that it will become saturated in the cooling solution long before any impurities will. But what can be done when a compound does not have the necessary physical properties for recrystallization or if it's not the dominant material in a mixture?

We really can't turn to recrystallization to obtain an effective separation. But there are other ways to isolate organic compounds from one another based on a phenomenon known as partitioning. Today we will talk about one of the most common classes of techniques used in this principle, extraction. But first, let's just think about partitioning itself in terms of something most of us can relate to, salad dressing.

Now, the contents of this bottle include water and olive oil. And we can clearly see the two ingredients, because they create two distinct phases. Since the oil and water are immiscible, or dissolve in one another, one simply floats on the other; in this case, the oil is above the water, because the oil is less dense. Now, also in this bottle are some other ingredients, including salt, and some herbs, and herb oils. Now, to simplify let's just say the recipe contains water and oil as solvents and salt and oregano, or the oil of oregano, as a solute. So the water and oil provide two different chemical environments in which the sodium and chloride ions from the salt and the organic carvacrol oil from oregano can dissolve.

Now knowing what we know about intermolecular forces and the role they play in solubility, we expect to find the ions of the salt preferentially dissolved in the high-polarity water, which can orient its bond dipoles to create favorable interactions with the dissolved ions. And of course, the low-polarity carvacrol from the herb oils dissolves in the low-polarity oil layer on top, because the larger oil molecules are more polarizable and can form dispersion forces which solvate the carvacrol.

It's exactly because of this separation that the proper blend of flavors is only obtained when we vigorously shake the bottle just before use. So to pour just the top layer would be to put only the herb oils on to my salad, while, to pour just the lower layer would only add the salt, acid, and water. So when we shake the dressing, we're getting ready for lunch. But when we don't shake the dressing, but rather, deliberately, allow the two phases to stay separate, dispensing each individually, we're doing a very simple form of liquid-liquid extraction.

Now, in our first example in this lecture, I used a system in which we assumed that all the particular solute accumulated in one of the two phases available to it. In part, I was able to make this assumption because of how drastically different sodium chloride and carvacrol are, being ionic and organic compounds respectively. But the truth for most organic molecules in a system like this isn't so simple. Many of them have an appreciable solubility in both water and a certain organic solvent. And when this is the case, we can expect a solute to establish a set of equilibrium concentrations in both solvents, based upon its relative solubility in each of them. We call this affect partitioning, and the equilibrium constant which governs it, is a partitioning coefficient. Partitioning coefficients are easily calculated from the maximum solubility of the solute in each of the two solvents. They give a numerical value, which allows us to predict how much of each solute will be in each layer at equilibrium.

So, let's take a look at a partitioning event as it happens. In this case, I'm going to use 1 butanol and water as solvents, and some blue-colored food dye as a solute. Now we can be pretty sure that this fairly polar food dye would rather be dissolved in the water layer, which is on the bottom of these two funnels. And yet, it's dissolved in the top layer right now, which is n-butanol, all a less polar solvent. It's up there now because I very carefully and very deliberately added it to the top without disturbing the lower layer, so that we could see the conditions before I actually perform a liquid-liquid extraction.

Now, the glassware that I'm using today are called separatory funnels, and their design is very specific for this purpose. They contain a ground glass stopper, which leads to a port on top, so that I can add my liquids to the separatory funnel itself, which is sort of a pear-shaped funnel, which leads to a Teflon stopcock, and that Teflon stopcock can be used to open the bottom of the funnel, draining the liquid through the stem. So I can add liquid to the top, but I can remove liquids from the bottom.

You can see I've already charged them very carefully with my two different solutions. So at the moment, my food dye dissolved in n-butanol is on top, because the density of n-butanol is about 80% out of water. And I have some pure water underneath. But that food dye is a very polar organic molecule, and it really would much rather be in the water than in the n-butanol. So why isn't it there already? Well, it's not there already because it needs a little bit of assistance. You see, partitioning takes place very, very slowly when we rely on nothing more than random molecular motion and a very narrow interface between the two solvents to allow them to jump from one side to the other.

They need a little bit of assistance from me in the form of agitation. So what I'm going to do next is take one of my two separatory funnels and agitate the mixture, creating many inclusions of one solvent in the other, thereby drastically increasing the surface area contact. Now, when I do this, it's going to take about a moment for everything to come back to equilibrium and for my two phases to separate again. But when they do, I think we'll see something very different than what we have now.

So let's take this separatory funnel, and again, agitate a bit to create some inclusions of one solvent in the other, periodically venting to make sure there's no pressure building up; we don't want that to happen. Then I'm going to return it to the stand. Now, if we look closely, we can already start to see the two phases separating. And that's not too surprising because n-butanol and water are immiscible in one another. But, what is surprising is, notice the different colors of the layers. Clearly, the food dye is accumulating in the lower layer. That's because it's more soluble overall in water than it is in n-butanol.

So I have affected a liquid-liquid extraction here. I've removed some, but not quite all, of the dye molecules from my n-butanol. Now, at this point, if it were my goal to retrieve those molecules, all I need is a beaker. Remove my stopper so that I don't have a reduction in pressure in the head space, and I can simply drain my aqueous layer through the stem and isolate the dye molecules and aqueous solution from my n-butanol. I think that's probably about enough. We can stop that.

Now, this is where the conical shape of the flask comes into play. If I watch my interface as I drain this, let's go for a little bit more here. If you watch the interface as I drain it, as it reaches the narrow portion of the stem, I can very carefully, very precisely cut the flow just as that reaches the stopcock, thereby collecting the maximum amount of my aqueous solution. So the dye that was in the n-butanol is now in water.

So, you probably noticed there that, even though most of that dye made its way into the aqueous layer, that a little bit of it remained in the organic n-butanol layer. So there was an equilibrium of concentrations there, and this equilibrium is governed by a constant which is equal to the maximum solubility of that dye in the aqueous solvent divided by the maximum solubility in the organic solvent. So that's what we call the partitioning coefficient.

So in this case, my dye had a very large partitioning coefficient, so I managed to extract most of it into the aqueous layer. But this isn't always the case for organic compounds. Sometimes they're pretty fussy about going into aqueous solutions. So, if I want to extract an organic compound from an organic solution into water, I'm going to have to play some tricks. And one of the best tricks we have is altering the pH of the aqueous layer.

So let's look at a representative organic molecule and start to think about the factors that affect its solubility in aqueous solution. For my demonstration, I'm going to use 2-naphthol. This is an aromatic alcohol with a pK_a of about 10. So, this hydroxyl group here is slightly acidic because of the resonance stabilization afforded by the benzene or the aromatic ring that it's attached to. Now we have to think back to our lecture on acids and bases and the Henderson Hasselbalch equation, which gives us a way to relate the amount of a weak acid, which is deprotonated, with that which is protonated as a function of the pH and of the pK_a of that compound.

If we do a little bit of rearranging and use the Henderson Hasselbalch equation to create a plot of the percent ionized in aqueous solution as a function of the pH of that aqueous solution, we get a trace that looks something like this. Notice the very steep transition that takes place right around the pK_a of the compound. So, in this case, if I were to attempt to dissolve napthol in water at a pH of about, let's say, 6, it's not going to dissolve as well as it could. And the reason for this is that at a pH of about 6, none of the dissolved napthol is going to be ionized, or practically none of it will be ionized. So if I were to pour some liquid into a beaker, adjust that liquid to pH 6, and attempt to dissolve napthol, only a very small amount will actually go into solution. Remember, partitioning coefficients are dictated by solubility. So if I can change its solubility, I can change its partitioning coefficient. Now, to do that, all I have to do is adjust the pH of my aqueous solution so that I'm on the other side of the pK_a.

So let's do that now. If I move over to a pH that's greater than 12, let's say 14. Four pH units above the pK_a, well in excess of what I would need. I'm not dissolving napthol anymore. I'm dissolving napatholate. I'm dissolving the ionized version of my weak acid. Now, polar water interact much better with ions than it does with neutral organics. And that means that these napatholate ions have a much higher solubility. So, by altering its solubility in water by adjusting the pH, I'm changing its partitioning behavior when I try to run a liquid-liquid extraction.

So let's do that now. Here's my napthol, but I've drawn two different equilibria. And the reason I've done this is that we now know that, depending upon the pH of an aqueous layer, I'll either be dealing with the equilibrium between 2-naphthol in the organic and 2-naphthol in the aqueous layer, or, between 2-naphthol in the organic and 2-naphtholate in the aqueous layer. Let's see how that affects the extraction properties by looking at a mock-extraction funnel here. I've got my Henderson Hasselbalch plot with my transition at the pK_a of my compound. And as long as I'm two pH units above or below that position, I'm in a situation where I can consider it to be either all ionized or all unionized.

So let's zoom in on this extraction funnel and see what happens. So here I am at the interface with an organic layer, let's say, ether, and an aqueous layer, which I've adjusted to a low pH, less than 8. That means that I'm dealing with an equilibrium between 2-naphthol in the organic layer and 2-naphthol in the aqueous layer. So at this point, I'm dealing with these neutral organic compounds, and they're not terribly soluble. So, at best, I could hope that a small fraction of them is going to move down here into the aqueous layer. I'm not going to be able to wash very much of it away in this fashion, so, my extraction is sub optimal at this point.

But what happens if I change the pH of that aqueous layer? Let's move to the other side of the Henderson Hasselbalch plot to where our pH is greater than 12, more than two units above the pK_a. Well, now I'm dealing with naptholate, and its solubility is higher, so its partitioning coefficient is greater. So simply, by carefully choosing the pH of my aqueous layer, I can make my extraction work better, or worse, depending on my goals.

So I can affect a different partitioning coefficient on a compound, at least if it's titratable, simply by adjusting the pH of the aqueous layer that I use. So, how can I use that to my advantage when I'm trying to extract one organic compound from another? And the answer to this lies in the fact that most organic compounds have slightly different pK_a values from one another and slightly different acid-base properties.

For example, we've already considered naphthol and naptholate, but take a look at naphthoic acid. Now, this is a compound that's very similar to naphthol, yet its pK_a is about 4, rather than 10. So if I were to plot it's Henderson Hasselbalch equation, I'd get a plot that looks something like this, in which I have a similar shape, but the transition is taking place at a much lower pH. So I can solubilize naphthoic acid in water much more easily; I only have to have a pH of around 6 or higher. Now, compare that to the equilibrium we just looked at, naphthol and naphtholate, where the pK_a is 10. It's Henderson Hasselbalch plot has a transition at a much higher pH. So that's a difference that I can use.

If we take it one step further and consider a compound which is not acidic, but is basic, for example, naphthylamine. Now, naphthylamine actually picks up a proton and becomes charged to form naphthylammonium ion, which has a pK_a of around 4.6, but because now naphthylammonium, the conjugate acid, is charged, not only will my Henderson Hasselbalch plot shift, but it will reverse on the vertical axis, and I'll have more ionized at a lower pH. Again, something I can exploit—difference in how they behave based upon the pH of the extraction solution. So again, our acidic compounds are charged when they're at high pH, but our basic compound is charged at low pH.

So now that we've laid this groundwork, we're ready to actually perform an extraction. Let's assume that we needed to separate this napthoic acid from 2-naphthol, two compounds whose pK_a and acid-based behavior we've already characterized pretty well. Their two Henderson Hasselbalch plots are kind of similar in shape, but they have a different transitional pH. So you can already begin to see, if you look carefully here, there is a region of the plot where they're behaving very differently from one another, but there are also regions of the plot where the behave very similarly to one another. And if we're going to separate them, our goal is to be in that zone where they're different.

So if we were to take a look at an extraction taking place with an organic layer on top and a very acidic aqueous layer on the bottom, what would we see? Well we would see a situation where neither of them is particularly ionized in the aqueous layer, and therefore, their solubilities are much greater up here in the organics. I've color coded them a little bit; I've shaded them in green and orange so they'll be easier to tell apart from one another. You'll see that the orange is my napthoic acid and that the greens are my napthol.

So, let's move along our Henderson Hasselbalch plot to a different pH and see if we can do better. If we move up to about neutral pH, around pH 7, take a look at what's happened here. I have reached a point of my Henderson Hasselbalch where there's a huge difference in the amount present as ions in the aqueous solution. My napthoic acid is completely deprotonated when in water, but my napthol is completely neutral. So their solubilities are very different, which leads to a situation where one partitions into the water as its conjugate base. The other remains up here in the organic layer as its conjugate acid. So if I were to attempt an extraction with an aqueous layer of this pH, separating the two is as simple as opening the stopcock at the bottom of the funnel.

But if I go too far, and I raise my pH all the way to, say, 12 or higher, I reach a point at which, not only is my napthoic acid going to be very water soluble, but so is my napthol, because now it can dissolve as naptholate. So at this point, I will have a preponderance of both compounds accumulating in the aqueous layer. My extraction will have been compromised.

And as a final example, let's take a look at the difference between weak acids and weak bases. For example, the extraction of napthol and naphthylamine. So if I had a physical mixture of these two and wanted to separate them using this technique, how would that be different from the extraction of two weak acids, one from the other, like we just looked at? Well remember, the Henderson Hasselbalch plot for my naphthylamine is reversed; it's flipped over on the vertical axis, so it's more ionized at low pHs and less ionized at high pHs. So what does that mean for my extraction?

Well, what that means is that if I start my extraction with a very acidic aqueous layer, I expect that my naphthylamine will actually be naphthylammonium ion, and therefore, very soluble in the water. So I'll start at a very low pH with an effective extraction, which places the naphthylamine in the aqueous layer and places the naphthol in the organic layer. But as I raise my pH, I run into a problem again. If I try this extraction with neutral water, I have a situation where my naphthylammonium is not soluble in water, because it's now naphthylamine. And it's going to migrate preferentially into the organic layer where the naphthol already is. So, by using a neutral aqueous solution, I'm not going to extract anything, at least not to any appreciable extent. Finally, if I continue, raise the pH even farther. I reach a situation where now my now naphthol can become deprotonated in the aqueous layer, leaving behind the naphthylamine. I have a good extraction once again, but this time, I've got my naphthol accumulating in the aqueous layer as opposed to the naphthylamine that I had at the beginning.

So one of the benefits of trying to extract an acid from a base is that we can choose which layer each will move into. I could put my basic compound in the aqueous layer by using an acidic aqueous solution, or, I can put my acidic compound in by using a basic. And of course, the drawback is, I have to use a lot of acid or base to get it done.

Now, so far in this lecture, we've thought of partitioning in terms of two immiscible liquid phases. But partitioning can also take place between any two distinct phases. Some other examples are solid liquid, or liquid-gas systems. So naturally, it should be possible to conduct extractions using these kinds of systems as well. Now, probably the most obvious example of an extraction using solid-liquid partitioning is the process of brewing a cup of tea.

As you add hot water to a tea leaf, certain compounds are going to dissolve in water better than others. In this case, the large and insoluble material making up the tea leaves can interact with, say, caffeine, polyphenols, and other compounds which would otherwise be soluble in cold water. But we all know from experience that steeping tea in cold water leads to a very weak beverage. So, just as we can manipulate the partitioning coefficient in a liquid-liquid extraction with pH, so we can manipulate the partitioning coefficients of compounds in solid-liquid systems using temperature, and tea connoisseurs will understand this well.

Green tea, it's famous for its delicate flavors and aromas, but green tea also contains particularly high levels of tannins, a polyphenolic class of compounds with a bitter taste and a very dry-mouth feel. So the goal in preparing this delightful beverage is temperature control. Fine green teas require careful attention to brewing temperature, as water just the below its boiling point partitions the pleasant and aroma-granting compounds into that water effectively. But water too close to boiling will separate the polyphenolic compounds, which have a more bitter taste, ruining the otherwise enjoyable experience of a well-crafted green tea.

Just take a look at the pleasant-tasting amino acid theanine, which makes up a sizable portion of the amino acid content of green tea. Now, take a look at tannic acid, a common polyphenol with a bitter, dry taste. Is it any wonder why theanine can be liberated from the hydroxyl-rich cellulose fibers of tea leaves more easily? All those hydrogen bonders, it's obvious. It simply clings on tighter. So a well-made green tea is a textbook example of a solid-liquid chemical extraction, selectively partitioning one type of compound into the liquid, while leaving the other behind adhere to the solid matrix. So if you're a green-tea drinker, I congratulate you on being a great bench chemist as well.

So let's review what we've covered in this lecture. We discussed the concept of immiscible liquids and how solvents at very low polarity tend to be immiscible in higher polarity solvents like water.

We talked about how organic solutes have a finite solubility in each solvent, and how we can use this finite solubility to calculate what's known as a partitioning coefficient. We then saw how the partition coefficient can be used to predict the relative amount of a given compound which will be found in each solvent at equilibrium.

Next we took a look at how the pH of water can affect the total solubility of titratable organic compounds in that solvent and how that can drastically change partitioning coefficients predictably and often usefully.

Finally we took a look at solid-liquid extractions in which partitioning occurs not between two immiscible liquid phases, but between being dissolved in solvent and being adhered to an insoluble solid. And we saw how this technique is no different than the quest for the perfect cup of green tea.

But our discussion of partitioning is only half over. We still need to consider one last caveat. See, this lecture's discussion focused on partitioning between two phases which are at rest. Next time, we're going to set one of those phases in motion, opening up the discussion to one of the most powerful separation techniques ever invented—chromatography. I'll see you then.

Purifying by Chromatography
Lecture 35

This lecture will explore the last topic on the science of separations. Instead of focusing on partitioning between two phases that are at rest, in this lecture, one of these phases will be set in motion, opening up the discussion to one of the most powerful separation techniques ever invented: chromatography. When properly applied, chromatography allows us to isolate almost anything we can imagine. From recrystallization, to distillation, to liquid extractions, to chromatography, there is a solution for nearly any separation problem that can come up in the lab.

Chromatography

- Mikhail Tsvet, the inventor of chromatography, was educated in Switzerland, where he received his Ph.D. in 1896. Shortly after this, however, he found himself in Russia, where his foreign credential nearly marginalized him. Tsvet's non-Russian credential was not recognized by the national establishment, prompting him to undertake a second Ph.D. program to become a functional scientist in the Russian system. It was the topic of this second Ph.D. that earned him immortality in the science of separation.

- Tsvet became interested in cell physiology during his first Ph.D. program and wanted to continue working with natural plant products, trying to understand the chemistry that drives their unique biology. It was during experiments with the ubiquitous plant pigment chlorophyll that he made the observation that would forever change his life and the science of separations.

- Tsvet knew that isolated chlorophyll could be easily dissolved in an organic solvent known as petroleum ether. However, when he attempted to extract the pigments from the leaves of plants, he noted that it was very difficult to dissolve. Even after grinding and tearing the leaves to expose more surface area and break open cells, the distinctive, dark green pigment simply wouldn't cooperate.

- His conclusion was that the chlorophyll pigment must be adhered to the solid plant matter through intermolecular forces, reducing its solubility in the solvent. Using this idea as a springboard for his research, Tsvet tried adhering chlorophyll to different solid surfaces, then washing it away with various solvents by allowing those solvents to flow through the solid matrix.

- He found that not only did chlorophyll migrate at different rates in different systems, but also that mixtures of pigments would many times separate in space because of the differing attractive forces at work between them and the two phases. This created an array of colors on Tsvet's column, prompting him to call his new technique chromatography, from the Greek words meaning "color" and "writing."

- Where Tsvet's creation really differed from simple liquid-liquid or solid-liquid extraction is that he added motion to the equation. By allowing one of the phases to move across the other, the partitioning of a solute between heterogeneous phases can be used to move compounds across the stationary phase at varying rates. The more time a particular compound spends partitioned into the mobile phase, the faster it moves.

Thin-Layer Chromatography and Paper Chromatography
- Chromatography has advanced considerably over the past century. Stationary phases like calcium carbonate and sugar from Tsvet's experiments have been replaced by a number of superior options. The first of these was actually paper.

- Paper consists of cellulose, a long polymer consisting mostly of interconnected glucose units. We already know that these glucose units have many hydroxyls that are available to interact with polar or hydrogen bonding compounds.

- English chemists Archer J. P. Martin and Richard L. M. Synge developed a method of chromatography using paper and organic solvents as stationary and mobile phases, respectively. In 1942, they published a paper outlining their technique but, more importantly, discussing the fundamentals of partitioning as it applied to chromatographic systems.

- This technique proved extremely useful in the identification of small organic molecules, including amino acids like glycine and alanine. This method was so influential that Martin and Synge received the 1952 Nobel Prize in Chemistry for this concept. A year later, young Stanley Miller used this method to verify the presence of those amino acids in his now-famous primitive Earth experiment.

- More recently, advances in material manufacturing have made quick identifications like Miller's even easier and more accurate. For example, microporous silica can now be manufactured. Silica is a fantastic stationary phase, because even though its formula is SiO_2, at its surface are an array of silenol groups, or $SiOH$. These groups interact well with anything of high polarity or hydrogen-bonding ability. Because silica can be manufactured with tremendous surface-area-to-volume ratios, more compounds can be separated with greater efficacy or on larger scales.

- For example, instead of the everyday paper used by Martin and Miller, modern organic chemistry researchers often use a thin-layer chromatography plate with a plastic backing with just a 200-micron-thick layer of silica bound to it. That is just about the width of a human hair, but that small amount of silica has a surface area of hundreds of square meters because of the extremely small size and porosity of the silica particles.

- In the technique of thin-layer chromatography, a small amount of the compound to be analyzed is spotted onto the plate and allowed to dry. Once the spot is dried, it is placed into a developing chamber containing a thin pool of the selected mobile phase.

- As the mobile phase wicks up through the plate by capillary action, the different compounds in the sample move at different rates. Compounds that interact more strongly with the polar silica move a shorter distance, while those that interact better with the lower-polarity mobile phase move a greater distance.

- At the completion of the experiment, we measure the distance traveled by the sample spot as a fraction of the distance traveled by the mobile phase and report this number as the retention factor (R_f) value for that sample in that particular system. This value gives us a semiquantitative way to describe simple chromatographic mobility.

Column Chromatography

- Thin-layer chromatography may be a versatile, quick, low-cost way to observe the chromatographic mobility of a compound, but working with such small quantities makes collection of a meaningful sample of the compound difficult.

- In order to collect a sample that can be used as a raw material, drug, analytical sample, or synthetic intermediate, we need to increase the scale of the experiment.

- This is frequently done in the lab using a technique known as column chromatography. In column chromatography, we abandon the thin layer of stationary phase on a backing for a column. Bringing the third dimension into play means that a 2-centimeter-wide column can separate about 8000 times as much sample as a TLC lane 200 microns thick.

- In this rather simple technique, we use a glass column with a Teflon stopcock at the base. The neck of the column is plugged with a piece of glass wool and filled with sand to provide a level layer onto which we can build a column of silica that is then saturated with the mobile phase.

- Then, we drain the mobile phase through the stopcock to expose the top of the column and gently add a narrow, concentrated band of the compound we want to analyze. We then drain that to get it in contact with the silica gel before topping it off with more mobile phase.

- After loading the column, we can open the stopcock and let it run. As the mobile phase moves downward under the force of gravity, the compounds again separate, but this time in large enough quantity that we can collect a band consisting of just one component. This fraction is now ready to be worked with. We can recover the solute by rotary evaporation, liquid extraction, or another technique, and then we are ready to work with the purified dye material.

Advanced Chromatography: HPLC

- In recent decades, chromatography has undergone a virtual explosion of advancements, leading to techniques involving ultramicroscopic silica particles with such small pores that powerful pumps must be used to push solvent through them at high pressure in a technique called high-performance liquid chromatography (HPLC).

- The extremely high surface area of the HPLC column packing allows very precise separations, but it also requires a closed system consisting of a steel column to be used so that it can resist the pressure applied by the pumps. So, it is impossible to load compounds in the same way as traditional chromatography, because the system is sealed. Similarly, it is impossible to see even colored compounds as they move through the system.

- The loading problem is solved with a device called an injection loop, which consists of a manifold with two separate loops made of a pressure-resistant tubing. When the valve handle is rotated, one loop is in line with the flowing mobile phase and the other is in line with a special injection septum.

- The sample is pushed into the open loop using a syringe, and then the handle is turned, placing the injection loop in line with the mobile phase, thereby introducing the sample into the chromatographic system without ever opening it and losing pressure.

- Because columns need to be packed into stainless steel cases, it becomes impossible to monitor a run with our eyes, even if our compounds are visible. So, HPLC systems also have a detecting system that is usually something like a simple spectrophotometer flow cell. The simplest example of this is a UV-visible detection system, in which a specific wavelength of light is aimed so that it passes through the eluting solvent, striking a detector.

- As the sample molecules move out of the column and through the detector cell, they absorb the light, leading to a reduced intensity at the detector. If we plot the observed absorbance as a function of time, starting with the injection of the sample at zero minutes, we can create what is called a chromatogram, or a graphical representation of the separation taking place.

Gas Chromatography

- There are many more chromatographic methods available to the modern chemist, including gas chromatography (GC). Archer J. P. Martin is the name most commonly associated with the invention of GC. Martin is actually most famous for his invention of paper chromatography, the technique used by Stanley Miller to detect the amino acids in his primordial concoction in Harold Urey's lab.

- Martin explored ways in which partitioning could be exploited to separate organic compounds faster and more effectively. It was around the time of his Nobel Prize that he hit on another great concept. Partitioning involves the motion of molecules from one phase to another, so why confine this methodology to transitions between adhered solid states and dissolved liquid states? After all, molecules move faster in gasses and slowest of all in solids. It stands to reason that molecules could switch from phase to phase more quickly if the gas phase were somehow included in the experiment.

Gas Chromatography

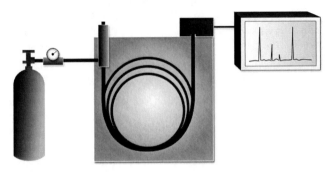

- Martin wondered whether a form of chromatography could be developed using liquid as the stationary phase and gas as the mobile phase. His idea proved viable. He demonstrated that separation could be accomplished with extreme speed and precision using a dense liquid phase and a gas like helium as the mobile phase. By slowly heating a column filled with the liquid stationary phase, through which a carrier gas like helium or nitrogen is flowing, compounds are driven off one by one in order of decreasing volatility.

- His new method allowed faster separation with far less material and has become a staple technique in forensic and analytical labs all around the world. Using various detection methods, GC can be used to analyze practically anything that will vaporize.

Suggested Reading

Scott, *Techniques and Practice of Chromatography*.

Wade, *Organic Chemistry*, 5.6.

Williamson, *Organic Experiments*, Chaps. 9 and 10.

1. What distinguishes chromatography systems from the liquid-liquid extraction systems discussed in the previous lecture?

2. If a mobile phase contains an aqueous component, how is the pH of that component expected to affect the mobility of basic or acidic compounds?

3. How will the chromatography of compounds change when a very low-polarity stationary phase is used in place of the very polar silica?

Purifying by Chromatography
Lecture 35—Transcript

Last time, we investigated the phenomenon of partitioning and how it can be exploited to separate compounds based on their relative affinities for two distinct chemical environments. Specifically we looked at liquid-liquid systems and solid-liquid systems. In this lecture we're going to explore one of the most powerful separation techniques ever developed, one which, when properly applied, allows us to isolate almost anything we can imagine. It's also a bit near and dear to my heart, because I spent two and a half years of my life chained to a lab bench analyzing pharmaceutical products with one of these techniques. You might say that in that way it was chromatography which inspired me to return to school for my Ph.D.

Our brief look at chromatography starts with its inventor, Mikhail Tswett. Tswett was educated in Switzerland, where he received his Ph.D. in 1896. Shortly after this, however, he found himself in Russia facing a similar fate to that of Zaitsev, whose foreign credential had nearly marginalized him in the Russian system. Tswett's non-Russian credential was not recognized by the national establishment, prompting him to undertake a second Ph.D. program to become a functional scientist in the Russian system. I can only imagine the sinking feeling he must have had when he first realized that he would have to earn his Ph.D. all over again. In retrospect, however, it may have been the single greatest thing that ever happened to him, since it was the topic of his second Ph.D. which earned him immortality in the science of separation.

You see, Tswett became interested in cell physiology during his first Ph.D. program, and he wanted to continue working with natural plant products, trying to understand the chemistry which drives their unique biology. It was during experiments with the ubiquitous plant pigment chlorophyll that he made the observation which would forever change his life and the science of separations.

Tswett knew that isolated chlorophyll could be easily dissolved in an organic solvent known as petroleum ether, yet, when he attempted to extract the pigments from the leaves of plants, he noted that it was very difficult to dissolve in the same solvent. Even after grinding and tearing the leaves to expose more surface area and break open cells, the distinctive dark green pigment simply wouldn't cooperate.

His conclusion was that the chlorophyll pigment must be adhered to the solid plant matter through intermolecular forces, reducing its solubility in the petroleum-ether solvent. Using this idea as a springboard for his research, Tswett tried adhering chlorophyll to different solid surfaces and then washing it away with various solvents by allowing those solvents to flow through the solid matrix to which he had adhered the pigment.

He found that not only did chlorophyll migrate at different rates in different systems, but also that mixtures of pigments would many times separate in space, because of the differing attractive forces at work between them and the two phases. This created an array of colors on Tswett's column, prompting him to dub his new technique chromatography, from the Greek words meaning color and writing.

Where Tswett's creation really differed from simple liquid-liquid or solid-liquid extraction is that he added motion to the equation. By allowing one of the phases to move across the other, the partitioning of a solute between heterogeneous phases can be used to move compounds across the stationary phase, but at varying rates. Obviously, the more time a particular compound spends partitioned onto the mobile phase or into the mobile phase, the faster it moves. Indeed, the modern definition of chromatography is, a separation technique, which relies on the differing affinity of a substance for a stationary phase and a mobile phase.

We can very closely duplicate the results of some of Tswett's experiments using a simple piece of chalk, some markers, and a solvent that's readily available to us, like ethanol. So, what I'm going to do now is I'm going to take that piece of chalk, to which I've applied a little bit of some pigments from a marker, and I'm going to use ethanol as a mobile phase to cause those dyes and pigments to migrate, but at different speeds.

So, ahead of time, I've already prepared my column, if you will, and as I promised, this is an experiment that's pretty easily done right in your own kitchen. This is nothing more than a piece of common sidewalk chalk. This is calcium carbonate, the exact same stationary phase used by Tswett in his original experiment. Now, on this piece of chalk I've placed bands of different colors simply by touching the tip of a marker to the chalk. So this chalk is now my stationary phase. It's going to stand perfectly still throughout the experiment.

I have a beaker here with a little bit of the solvent ethanol in it. Now, I'm using ethanol, instead of petroleum ether, because the pigments from the markers are actually much more polar molecules than the pigments that Tswett was separating. So I need a little bit more polar solvent to encourage this ink to migrate. What I'm going to do now is place my chalk into this beaker containing a very shallow pool of my mobile phase ethanol and allow it to migrate up through the chalk through, essentially, capillary action. So my chalk will absorb the solvent, and chromatography will take place. So let's start that now.

I'm going to very carefully balance my chalk here and place a piece of aluminum foil over the top in an attempt to discourage evaporation of my mobile phase while separation takes place. So if we wait for just a moment, we should already be able to see the chalk is becoming darker, as the solvent moves up toward the ink. So before long, it's going to pass over my bands of different colors and begin to separate those molecules, based upon chromatography.

Of course chromatography has advanced considerably over the past century. The stationary phases, like calcium carbonate and sugar from Tswett's experiments have been replaced by a number of superior options. The first of these was actually paper. You see, paper consists of cellulose, a long polymer consisting mostly of interconnected glucose units. And we already know that these glucose units have many, many hydroxyls which are available to interact with polar or hydrogen-bonding compounds.

Two English chemists by the name of Archer J. P. Martin and Richard Synge, developed a method of chromatography using paper and organic solvents as stationary and mobile phases, respectively. In 1942 they published a paper outlining this technique, but more importantly, it discussed the fundamentals of partitioning as it applied to chromatographic systems. This technique proved extremely useful in the identification of small organic molecules, including amino acids like glycine and alanine. This method was so influential that Martin and Synge received the 1952 Nobel Prize in Chemistry for this concept. Not only did these two receive the prize for their work, but a further validation came just a year later, when a young scientist named Stanley Miller used this method to verify the presence of amino acids in his now-famous primitive Earth experiment.

More recently, advances in material manufacturing have made quick identifications, like Miller's, even easier and more accurate. For example, micro-porous silica can now be manufactured. See, silica is a fantastic stationary phase, because, even though it's formula is SiO_2, at its surface are an array of silanol groups or SiOH motifs. These groups will interact well with anything of high polarity or good hydrogen-bonding ability. And because silica can be manufactured with a tremendous amount of surface area per volume ratio, more compound can be separated with greater efficiency, or on larger scales.

For example, instead of the everyday paper used by Martin and Miller, modern organic chemistry researchers often use a thin-layer chromatography plate with a plastic backing and just a 200-micron-thick layer of silica bound to it. That's just about the width of a human hair. But that small amount of silica has a surface area of literally hundreds of square meters, because the extremely small size and porosity of the silica particles gives it this very, very large surface-area-to-volume ratio.

In the technique of thin-layer chromatography, a small amount of compound to be analyzed is spotted onto the plate as a solution, and it's allowed to dry. So once that spot is dried, it's placed into a developing chamber which contains a thin pool of the selected mobile phase, which will move up the plate in the very same way that our mobile phase moved up the chalk in our previous demonstration.

So our chalk has been developing for a little while. And even though the solvent has only run a very small distance, it's already very easy to see what's been going on here. So let's take a closer look. I'm going to remove my stationary phase along with the adhered pigments. I'll move my ethanol to the side. Now, if I turn this chalk you can see very clearly that different colored markers contain different colored pigments. And not only that, but those different colored pigments have traveled different distances.

Now, since calcium carbonate is a particularly polar substance and ethanol is an organic solvent, I can say with relative certainty, for example, that the yellow dye to produce the color of the yellow marker has a higher polarity overall than do the molecules that make up, say, this blue dye that was included in the green marker. And I can make this sort of qualitative assessment of the polarity of the different dyes within simply by looking at how they move during my chromatography experiment.

So let's take a look at a thin-layer chromatography experiment, just like we did our chalk experiment, which simulated the work of Tswett. Now this time I'm going to put on some gloves, just because I'd like to hold up for you and show you what a thin layer chromatography plate actually looks like. Now they come in many sizes, and they come with many different stationary phases, but by far the most common uses silica. And it's backed in plastic, so the shiny side on the back here is actually plastic, on which chromatography would never happen, but this is the surface which contains the very, very thin layer of microscopic beads of silica. In fact, this jar contains loose silica, the likes of which would be adhered to the backing of the plate. So as you can see, it's extremely fine particle size, and that gives it its extremely large surface-area-to-volume ratio.

Now I have a small developing chamber here, which has alcohol in it, just as before. And I've created a plate, which I've spotted with some of the markers that we used in the previous experiment as well. And you're probably going to notice a marked difference in their chromatographic behavior, because, even though I'm using the same mobile phase, we're changing our stationary phase now from calcium carbonate to silica, both very polar, but in their own way. So we're going to see a drastically different separation pattern taking place. Nonetheless, the development process is essentially the same.

I'm going to take my plate onto which I placed three small spots, one of green, one of orange, and one of the red marker. And I spotted them just as

I did in my bars on the chalk stick in such a way that they don't submerge beneath the thin pool of mobile phase, because I don't want them to dissolve into all that mobile phase. So I'm going to place that inside very carefully. I'm going to cover that back up with my foil to discourage any evaporation of my mobile phase. And fairly quickly, we should be able to see that the alcohol mobile phase is making its way up the plate, once again. And just as before with the chalk, when it encounters the spots, they will begin to migrate at a rate which is consistent with their relative affinity for the mobile and stationary phases. So we'll come back to this and take a look at it in just a few minutes, when it's done

So the mobile phase has moved an appreciable distance up the plate, which is the stationary phase in my TLC experiment, so I'm ready to take a look now at the result. I'm going to remove my plate. Now, if I were in the lab, I would very quickly take a pencil and mark the lane. Because, as we'll see here briefly, over time, I expect that that alcohol mobile phase will begin to evaporate, and it won't be quite so clear as it is now where it traveled to. Now, it's very important for me to know that, because I can use the distances traveled by the spots relative to that travelled by the mobile phase to get some numbers that we'll talk about in a moment.

But first let's just take a look at the plate. So it's very obvious that the green marker actually contains some blue dye. Now we may not have guessed that originally, simply from looking at the marker itself, but the chromatography plate makes this fact, essentially, undisputable. Not only that, I can say that some of that blue dye moved very well with the mobile phase. And therefore, it didn't have a terribly high affinity for the stationary phase. In other words, it's moving that way because it's less polar than the others.

We can even see similar components in the red and orange markers but in slightly different proportions, and, of course, the orange having more yellow and the red having more of the darker pigments in it. So this is how TLC can be used to very quickly get qualitative assessments of compounds, polarities, and solubilities in different solvents, but also to give us a very quick and easy way to compare one lane to the next and identify common compounds in different samples.

Now since my solvent is still evaporating, I'll show you another one which I prepared earlier, in which the solvent has all but evaporated completely. And again, you can see, I've got basically the same result. And yet another very important property of this chromatography technique is that it's reproducible. So as long as I use the same stationary and mobile phases each time, I can expect the same relative mobilities for all the spots within a given sample.

So as the mobile phase wakes up through the plate by capillary action, different compounds in our samples were moving at different rates. Now, compounds which interact more strongly with the polar silica moved a shorter distance, while those which interacted better with the lower polarity mobile phase moved a greater distance. At the completion of the experiment, we can measure those distances traveled by the sample spots as a fraction of the distance traveled by the mobile phase. And we report that number as the RF value for that sample in our particular chromatographic system. RF stands for retention factor, and it gives us a semi-quantitative way to describe simple chromatographic mobility, like we just saw in our experiment.

Now, thin-layer chromatography may be a versatile, quick, low-cost way to observe the chromatographic mobility of a compound, but working with such small quantities makes collection of a meaningful sample of the compound somewhat difficult. In order to collect a sample which can be used as raw material, drug, an analytical sample, or a synthetic intermediate, we need to increase the scale of the experiment. This is frequently done in the lab using a technique known as column chromatography. In column chromatography, we abandon the thin layer of stationary phase on the backing in favor of a column. Now, this brings the third dimension into play. And this means that a two-centimeter wide column can separate about 8,000 times as much sample as a TLC lane, which is just 200 microns thick.

Now, we accomplish this using a device like this. It's a glass column, which contains a Teflon stop cock at the base and the stem, through which we can drain mobile phase. I place a small bit of glass, wool, or some other obstruction in the neck to prevent solids from moving through, then a nice, even-layered bed of sand onto which I would build a column of silica, again, using this finely, finely powdered silica material.

The key here is that because I'm using a column which has depth, instead of that extremely thin layer of silica, I'm using a very large bed of silica. In fact, the bed contains about 8,000 times the volume per unit length in a column this size, compared to the kind of TLC plates that we saw previously. So using the column, I can get a larger quantity of material. The only real catch here comes when we try to analyze the mobility. We have to remember that in our previous examples the mobile phase migrated from the bottom of the column to the top, but in this case, once we pack the column in that sample, we'd be adding mobile phase through the top and draining to the bottom, meaning that all the motion is now downward. So a common point of confusion is that even though the bands may appear to be reversed from the TLC experiment, they're in fact, if they are reversed, exactly what we would expect.

Now, the techniques that we've looked at in this lecture so far are just the very basic techniques, taught at the introductory level, though, still used frequently in research labs today. In recent decades, though, chromatography has undergone a virtual explosion of advancements, leading to techniques involving ultra-microscopic silica particles with such small pores that powerful pumps have to be used to push solvent through them at high pressures. We call this technique HPLC for high pressure liquid chromatography. The extremely high surface area of the contents of an HPLC column allows very precise separations, but it also requires a closed system consisting of a steel column to be used, so that it can resist the pressure applied by pumps, which move the solvent. So it's impossible to load compounds in the same way as traditional chromatography since the system is sealed. Similarly, it's impossible to see even the brightest colored compounds as they move through the system, because the column is encased in steel.

Now the loading problem is solved with the device called an injection loop. An injection loop consists of a manifold with two separate loops made of a pressure-resistant tubing. So, when the valve handle is rotated, one loop is in line with the flowing mobile phase, and the other is in line with a special injection septum. The sample is pushed into the open loop using a syringe. And when the handle is turned, places the injection loop in line with the mobile phase, thereby introducing the sample to the chromatographic system without ever opening it and losing pressure.

Of course, since columns need to be packed into stainless steel cases, it becomes impossible to monitor a run with my eyes, even if my compounds are absorbing the visible light. So HPLC systems also have a detecting system, which is usually something like a simple spectrophotometer flow cell. The simplest example of this is a UV-visible detection system, in which a specific wavelength of light is aimed so that it passes through the exiting solvent, striking a detector. As the sample molecules move out of the column and through the detector cell, they absorb the light, leading to a reduced intensity at the detector. So if we plot the observed absorbents as a function of time, starting with the injection of the sample at zero minutes, we can create what's called a chromatogram, or a graphical representation of the separation taking place.

There are many more chromatographic methods available to the modern chemist, certainly many more than we can survey in a single lecture. But no lecture on chromatography would be complete without the mention of gas chromatography, commonly referred to as GC. Archer J. P. Martin is the name most commonly associated with the invention of gas chromatography. Now, Martin is actually most famous for his invention of paper chromatography, the technique used by Stanley Miller to detect the amino acids in his primordial concoction made in Harold Urey's lab. Martin shared the 1952 Nobel Prize for that work, but like most Nobel Laureates, even the indisputable pinnacle of scientific honors was not enough to satisfy his curiosity. Martin continued to explore ways in which partitioning could be exploited to separate organic compounds faster and even more efficiently.

It was around the time of his Nobel Prize award that he hit on another great concept. Partitioning involves the motion of molecules from one phase to another. So, why confine this methodology to transitions between adhered solid states and dissolved liquid states? After all, molecules move faster in gases, and slowest of all in solids. So it stands to reason that molecules could switch from phase to phase much more quickly if the gas phase were somehow included in the experiment. So Martin wondered, could a form of chromatography be developed using liquid as the stationary phase and the gas as the mobile phase?

Martin's idea proved viable. He demonstrated that separation can be accomplished with extreme speed and precision using a dense, liquid phase and a gas, like helium, as the mobile phase. By slowly heating a column filled with a liquid stationary phase, through which a carrier gas, like helium or nitrogen is flowing, compounds are driven off one by one in order of decreasing volatility. His new method allowed faster separation with far less material and has become a staple technique in forensic and analytical labs all around the world. Using various detection methods, GC can be used to analyze practically anything that will vaporize.

In this lecture we explored our last topics on the science of separations. Specifically, we continued our investigation into partitioning, which is the equilibrium that compounds naturally establish between two distinct phases, based upon their relative affinity for those phases.

We took a look at the work of Mikhail Tswett, whose second Ph.D. dissertation gave rise to the science of chromatography, in which partitioning between a stationary and mobile phase leads to predictable rates of movement for compounds. This movement is based upon their relative affinity for the two phases.

We saw how this very basic principle plays out on a piece of blackboard chalk, much like Tswett's first experiment, and then again on a TLC plate, which is designed for small-scale characterizations and method development.

And finally, we talked about columns, which could allow us to collect enough sample to carry our work forward and perform additional reactions or testing.

Finally, we briefly mentioned some of the more powerful techniques in use today for special applications, from quality testing to medical research, and beyond.

We looked at HPLC, which uses extremely fine silica with very small pores to generate a huge surface area, giving us better separations but requiring powerful pumps operating at thousands of pounds per square inch to keep that mobile phase moving.

And we also saw how Archer J. P. Martin's second claim to fame, gas chromatography, operates by partitioning between the liquid and gas phases, rather than the traditional solid and liquid phases. And we discussed how this gives more partitioning events in a given unit of time, creating a fantastic resolution in a very short experiment.

From re-crystallization, to distillation, to liquid-liquid extractions and chromatography, there is truly a solution for nearly any separation problem that could possibly come up in the lab. We've come a long way, from atomic and molecular structure, to the synthesis of organic compounds, a host of identification techniques, and methods used for purification.

In the last 35 lectures we have managed to gather an understanding of some of the most basic tenets of organic chemistry. But we spent most of our time looking back on the history of this complex and powerful science, an absolute must if one is to gain the perspective to meaningfully look forward at the possibilities that organic chemistry holds for the future. So, the topic of our final lecture will be just that. We'll experience the joy of taking some of what we learned from the past 35 lectures, adding a dash of imagination, and trying to gaze into the future through the eyes of an organic chemist. I'll see you then.

The Future of Organic Chemistry
Lecture 36

From atomic and molecular structure, to the synthesis of organic compounds, to a host of identification techniques and methods used for purification, in the last 35 lectures you have gained an understanding of some of the most basic tenets of organic chemistry. In this final lecture, you will experience the joy of taking some of what you have learned throughout this course, adding a dash of imagination, and trying to gaze into the future—through the eyes of an organic chemist.

The Origins of Life

- Our understanding of the origins of life continues to evolve. Even as the scientific community begins to get a handle on just how complex biological systems can be—such as those driving DNA translation or protein structure and function—we still struggle with the simplest question of all: How did it all get started? Even now, evidence of exactly how the first carbon-containing compounds blinked into existence on Earth is scarce.

- Stanley Miller gave us some insight into how certain biological materials might form under the conditions of the primitive Earth's atmosphere 3 billion years ago, but there are those who believe that these molecules did not form on Earth at all—that life may have fallen to Earth from outer space.

- In 1969, a huge fireball rocketed across the Australian sky before separating into several pieces and finally crashing to Earth near the town of Murchison, Victoria. This meteorite strike is unusual because it was witnessed, so there is no debating its extraterrestrial origins; it was rather large, delivering more than 100 kilograms of material; and it appears to have carried with it a buffet of organic compounds.

- In 2010, careful chromatographic separation of the extracts analyzed by mass spectrometry and NMR spectroscopy revealed that the meteorite contained thousands or even tens of thousands of different small organic molecules encased within. The implication is that just like the meteorite, which is unquestionably an authentic space rock, those molecules that were trapped within its matrix must have fallen from space.

- Even more tantalizing evidence of extraterrestrial organics has been collected in recent decades as organizations like NASA launch sophisticated probes and telescopes like the Spitzer Space Telescope, which operated in the middle of the first decade of the 21st century. Spitzer found not only evidence of carbon, but specifically of sp^3-hybridized carbon.

Chirality

- Another curiosity that scientists are still trying to address is that all life on Earth that we know of uses l-amino acids, or left-handed amino acids, as the principle constituent of the proteins of life. That means that even though we may think of ourselves as being achiral at the macroscopic level, we are in fact not symmetrical when we look at ourselves through molecular eyes. At the molecular level, we are chiral.

- However, every reliable source of data on abiologically synthesized organics suggests that racemic mixtures of these compounds are created in natural processes. This leads us to two important questions. First, why l-amino acids? The molecular machinery that our bodies use to create the proteins and enzymes we need to live are chiral themselves, so it makes sense that we should use all of one handed amino acid or the other. So, clearly, nature had to make a choice early on: left-handed biochemistry or right-handed biochemistry?

- But is there any real difference between our biochemical world and its mirror image? Is there some quantum mechanical effect that we do not understand perfectly that dictated that choice, or was it just a cosmic coin toss that led us to be composed of l-amino acids instead of d-amino acids?

- This question was answered in the early 1990s, when the labs of Stephen Kent at The Scripps Research Institute used modified techniques pioneered by Bruce Merrifield to create a perfect mirror image of the protein HIV protease. He then tested it for activity against mirror images of its normal substrates, finding that the chemical activity of the right-handed protein was identical in every way to its left-handed version.

- This closed the book on the question of handedness in the chemistry of life. Racemic sources of material that led to our left-handed biochemistry should be equally capable of seeding right-handed biochemistry. So, if we ever do make contact with extraterrestrial, carbon-based life-forms, it would appear that there is a 50% chance that their biochemistry will be a perfect reflection of ours.

- Kent's research opened up a whole new vein of inquiry. Now that we know that d-amino acids can be used to create enzymes with every bit as much power to promote highly specialized chemistry, researchers are trying to develop new therapeutics made from d-amino acids.

Biomimetic Chemistry

- When humans sought to take to the air and fly, birds and their wings were an obvious inspiration for the design of early aircraft. Scientists call this kind of design strategy—one of observing the properties of natural systems and applying them to engineering—biomimicry.

© Library of Congress Prints and Photographs Division. LC-DIG-ppmsca-02546.

Biomimicry is evident in the designs for early aircraft wings.

- Similarly, as we learn more and more about the biomolecular world, scientists more often turn to the biological for inspiration in their designs of useful compounds. This practice of imitating the function of biological molecules is called biomimetic chemistry, and it may provide the springboard that we need to create small molecules with the exact chemical properties that we need to accomplish a number of tasks.

- Researchers at the University of Leeds have successfully synthesized what they call porphyrin cored hyperbranched polymers, which have oxygen-binding properties similar to hemoglobin but can be easily created in a lab. Someday, compounds like this one might offer a non-biological oxygen carrier for use in medical applications like surgery. An artificial blood like this would all but eliminate concerns over disease transmission or incompatible antibodies associated with human blood transfusions.

Synthetic Life
- Our ability to imitate life goes far beyond creating small, mimetic molecules. In 2010, scientists at the J. Craig Venter Institute reported that they had chemically synthesized a genome of over a million base pairs and then substituted that DNA for the native DNA in bacteria.

- In this proof-of-concept experiment, the researchers made just a few small changes between the natural DNA of the bacteria and their synthetic form, targeting regions of the genome that were known to act as structural support rather than those coding for specific proteins. Such a simple change was enough that it could be detected in the cells that had accepted the transplant without compromising the viability of the cells.

- So, the genetic deck was stacked in favor of success in this experiment. Still, this achievement will no doubt take us to the next level of biomolecular engineering, as it demonstrates clearly and effectively that the molecular machinery of life is modular and can be transplanted from one cell to another.

- This experiment shows that with enough understanding and cautious, dedicated effort, genetic material and the cellular machinery that uses it can be mixed, matched, and altered to produce any biochemistry we desire. Someday, we may even create entire cells from chemicals on a lab bench using organic reactions that are available to us today.

Carbon Sequestration

- Carbon-containing compounds contain a great deal of chemical energy, and whether we are burning fuels or using foods for respiration, a large amount of the organic material in the world is destined to become carbon dioxide. In addition, when humans and other creatures are done with their biomass for good, decomposition naturally releases much of their carbon as CO_2.

- Carbon dioxide acts as a greenhouse gas, absorbing radiated heat from the Earth's surface, leading to climate change when it is not properly balanced. This has led to a global movement to devise ways to store and use carbon dioxide in ways that prevent its release into the atmosphere.

- For example, the U.S. Department of Energy has estimated that 2.4 billion metric tons of industrially produced CO_2 could be stored by injecting it into subsurface structures like un-minable coal seams, where it can become adsorbed to the surface of the underground carbon.

- The concern is that there are other gasses already bound to the coal in seams like these, most notably methane. Those who advocate this method are betting on irreversible binding of carbon dioxide to subsurface structures like the carbon-rich coal deposits of coal seams—binding so strong that it can displace methane.

- There's a good chance that such an idea would work, but it depends on how well each environmental gas adsorbs to the stationary coal phase relative to its tendency to vaporize. In complex geological systems like these, the overall composition and chemical behavior of formations can be tricky to predict, so we won't know the best method for sequestering CO_2 until we try and see the results.

Molecular Engineering

- When we have completely exhausted the ability of our planet to provide for us, even with some assistance from science, the natural next step is that we look beyond the Earth for places to live, continue to expand our knowledge, and carry our species into the reaches of the solar system—and eventually the cosmos.

- To do so, at least to get started, will require that we take with us everything and everyone that a colony might need to get to and survive on a new celestial body. But that leads to a number of logistical questions, including how to get that much material off of the Earth and into space. With current rocket technology, it costs thousands or even tens of thousands of dollars to get just one pound of material into orbit around the Earth.

- There are many possible approaches to the challenge of finding a cost-effective way to escape the gravity of Earth. One very interesting proposal is the construction of a so-called space elevator, which would consist of a line that reaches from the equator of the Earth to an altitude of about 36,000 kilometers, the altitude at which a geostationary orbit is obtained.

- It is unlikely that any such structure could support its own weight. One proposed solution to this problem is to use a structure that is supported not by compression, but by tension—constructing a cable with a counterweight reaching so far into space that its center of mass is above 36,000 kilometers and whose overall length is 100,000 kilometers so that it is held taut under its own momentum as it turns with the Earth.

- Using a cable like this, a mechanical climber could simply move up the cable, taking its payload of people and supplies to the necessary altitude to escape Earth's gravity and reach outer space. Estimates are that a device like this, though profoundly expensive to build, once operating could lower the cost of delivering material into space by more than tenfold.

- In order to construct this modern marvel, a cable about 100,000 kilometers in length—more than twice the circumference of the Earth—would be produced. But to build a cable that long that can withstand the tension produced by its orbiting counterweight, we have to find the lightest, most durable material with the greatest tensile-strength-to-weight ratio ever developed. And the front-runner is carbon nanotubes.

- Under one proposal, nanotubes several meters in length could be woven together into a rope, creating the longest man-made object in human history and providing us with the cable needed to make the space elevator a reality.

- Researchers have reported that they have managed to create single nanotubes with lengths approaching the necessary mark needed to form the proposed cable. So, it would seem that soon the only thing standing in the way of a space elevator will be our own imagination—and about $20 billion.

Suggested Reading

Deamer and Szostak, *The Origins of Life*.

Edwards and Westling, *The Space Elevator*.

Questions to Consider

1. How might the small organic compounds found within the Murchison meteorite have formed?

2. Are there any other materials that rival the strength of carbon nanotubes that might be used in the construction of a so-called space elevator? Even once a cable material has been successfully developed, what other engineering challenges remain to be addressed?

The Future of Organic Chemistry
Lecture 36—Transcript

Welcome to our closing lecture on organic chemistry. The final lecture of this course is bittersweet for me. But it's also exciting for me to think about how much you've learned and that you are now ready to take that knowledge with you out into the world.

We've spent 35 lectures exploring the definition of organic chemistry, the theories upon which it is built, and many examples of how organic chemistry as a science has touched our lives and steered our history as a species. It's unlikely that Lavoisier and Wohler ever dreamed that simply by challenging historical ideas that they would help to start 300 years of experimentation, which would ultimately lead us to where we are today, at the precipice of understanding the origins of our own existence. The scientists who followed them carry that understanding forward in ways which have improved the lives of billions of people on the planet today and countless more who will come after them.

So, now that we understand and appreciate the contribution of these past researchers, I think we've earned the right to have a little bit of fun. Let's turn our attention forward and dream about the future that will be created by the scientists who stand on the shoulders of these giants who helped us understand the chemistry of carbon-based compounds. One of the most exciting frontiers in modern organic chemistry is its interface with other scientific disciplines, like biology, engineering, and Earth sciences. So why not start there and ask the question, how might this interface help us create a better world for ourselves in the coming decades?

If you'll pardon the pun, our understanding of the origins of life continues to evolve. Even as the scientific community begins to get a handle on just how complex biological systems can be, like those driving DNA translation or protein structure and function, we still struggle with the simplest question of all. How did it all get started? Even now, evidence of exactly how the first carbon containing compounds blinked into existence on Earth is scarce.

Stanley Miller gave us some insight into how certain biological materials might form under the conditions of the primitive Earth's atmosphere some three billion years ago. But there are those who believe that these molecules didn't form on Earth at all; there are those who think that life may have literally fallen to Earth from outer space.

In 1969, a huge fireball rocketed across the Australian sky before separating into several pieces and finally crashing to Earth near the town of Murchison, Victoria. This meteorite strike is unusual in three important ways. First, it was witnessed, so there's no debating its extraterrestrial origins. Second, it was rather large, delivering more than 100 kilograms of material to the surface of the Earth. And finally, the most fascinating characteristic of this meteorite is that it appears to have carried with it a buffet of organic compounds which is hard to believe.

As spectroscopic techniques continue to advance, the Murchison meteorite has been analyzed from time to time, always with new and interesting results. In 2010, for example, a group of researchers attempted to extract organic compounds from freshly broken samples of the Murchison. Now, careful chromatographic separation of the extracts analyzed by mass spectrometry and NMR spectroscopy techniques revealed that the meteorite contained not just a few, but thousands, or even tens of thousands, of different small organic molecules all encased within it.

Of course, the implication here is that just like the meteorite, which is unquestionably an authentic space rock, those molecules which were trapped within its matrix must have fallen from space. So this brings up new questions. If the Murchison results are accurate, just how did these compounds get there? Even more tantalizing evidence of extraterrestrial organics has been collected in recent decades as organizations like NASA launch sophisticated probes and telescopes, like the Spitzer Space Telescope. The Spitzer operated in the middle of the first decade of the 21st century. And it looked at interstellar space through infrared eyes, sensing absorptions caused by organic molecular vibrations occurring tens or even hundreds of thousands of light years away or more.

Perhaps not surprisingly in light of the Murchison meteorite results, Spitzer indeed found not only evidence of carbon, but specifically, of sp^3-hybridized carbon. A clear absorbance at 2,900 wave numbers was reported when the telescope was pointed at interstellar space. That's an absorption corresponding to the CH bond vibrations one would find in a typical organic compound. The implication? That organic carbon does indeed exist in abundance in regions of space which harbor little else. It would seem that even the void of deep space may harbor the seeds of life. So the more we search, the more we seem to find evidence that organics can spontaneously form not only on Earth, but elsewhere in the universe, potentially seeding the cosmos with life.

Yet another curiosity which scientists are still trying to address is this. All life on Earth that we know of uses L, or left-handed, amino acids as the principal constituent of the proteins of life. That means that, even though we may think of ourselves as being achiral at the macroscopic level, we are, in fact, not symmetrical when we look at ourselves through molecular eyes. For example, the amino acids in proteins making up the skin and muscle in your arms are all of one handedness. So if you could look at your right index finger and your left index finger with atomic resolution, they would, in fact, not be reflections of one another. So at the molecular level, you are chiral.

Yet, every reliable source of data on abiologically synthesized organics, from those formed in Miller's primitive Earth experiment to the interstellar stowaways on the Murchison meteorite, suggest that racemic mixtures of these compounds are created in natural processes. This leads us to two important questions. First, why L amino acids? The molecular machinery which our bodies use to create proteins and enzymes we need to live are chiral themselves, so it makes sense that we should use all of one handed amino acid or the other. So clearly, nature had to make a choice early on—left-handed biochemistry or right-handed biochemistry.

But is there any real difference between our biochemical world and its mirror image? Is there some quantum-mechanical effect that we don't understand perfectly which dictated that choice? Or, was it a cosmic coin toss that led us to be composed of L amino acids instead of D? This question was effectively answered in the early 1990s when the labs of Stephen Kent at Scripps

Research Institute created a right-handed version of a naturally-occurring left-handed protein, HIV-protease.

Kent used modified techniques pioneered by Bruce Merrifield to create a perfect mirror image of that protein. He then tested it for activity against mirror images of its normal substrates, finding that the chemical activity of his right-handed protein was identical in every way to the left-handed version. This closed the book on the question of handedness in the chemistry of life. Racemic sources of material which led to our left-handed biochemistry should be equally capable of seeding right-handed biochemistry. So if we ever do make contact with extraterrestrial carbon-based life forms, it would appear that there's a 50% chance that their biochemistry will be not exactly the same as our own, but rather, a perfect reflection of our own.

But Kent's research opened up a whole new vein of inquiry. Now that we know that D amino acids can be used to create enzymes with every bit as much power to promote highly specialized chemistry, researchers are trying to develop new therapeutics made from those D amino acids. It may be that someday, these mirror images, which previously had no use in biology, or little use in biology, will find use in the hands of humans who can not only rationally design them, but produce them in laboratories for use as medicines, chiral catalysts, and more.

When humans sought to take to the air and fly, birds and their wings were an obvious inspiration for the design of early aircraft. In fact, evidence of working bird-shaped gliders may date as far back as ancient Egypt. Now, scientists call this kind of design strategy—one of observing the properties of natural systems and applying them to engineering—biomimicry. And we've been doing this for a very, very long time.

And just as the designs of some planes, trains, and boats have been inspired by nature, as we learn more and more about the biomolecular world, scientists more and more turn to the biological for inspiration in their designs of useful compounds. This may one day help us to accomplish similar chemistry to the molecules of life but with smaller compounds, which are more easily obtained and whose properties are more easily adjusted and tuned than their larger biological counterparts.

We call this practice of imitating the function of biological molecules, biomimetic chemistry. And it may provide the springboard that we need to create small molecules with the exact chemical properties that we need to accomplish a number of tasks. Take the example of oxygen transport and storage. Our bodies and those of many other animals rely on hemoglobin to bind, transport, and release oxygen within our bodies so that our cells have the oxidizer that they need to carry out cellular respiration and give us life-sustaining chemical energy.

Hemoglobin consists of four proteins bound together in a quaternary structure which supports four organo-metallic heme molecules. It's here, at the iron atom within the heme, that oxygen binds when it is present in great abundance, like in the air that enters our lungs. Now, that oxygen is then released in environments which are lower in oxygen concentration, like the tissues of our body, where it's most needed.

The elegant balance of binding and releasing oxygen is mediated by the protein itself, which affects solubility and tunes the chemistry of the heme so that it grabs oxygen with striking efficiency in the lungs then transports it through the blood and releases it at just the right time as well. So proteins, in a way, are nature's organic chemists, tinkering with the properties of small molecules and asserting an influence over how they behave. Well, that sounds a lot like something that humans are learning to do for themselves, doesn't it?

So can we abandon the large protein and substitute the knowledge that we've gained over the past two centuries to create other molecules which can imitate the properties of hemoglobin? Researchers at the University of Leeds think so. They successfully synthesized what they called porphyrin-cored hyperbranched polymers, which have oxygen-binding properties very similar to hemoglobin but can be easily created in a laboratory.

Looking at the structure of a porphyrin-cored hyperbranched polymer, it's easy to see where the inspiration came from for this compound. It's also easy to see how someday compounds like this one might offer a non-biological oxygen carrier for use in medical applications, like surgery. And artificial blood like this would all but eliminate concerns over disease transmission or incompatible antibodies associated with human blood transfusions.

With available polymer arms, which could be modified to achieve almost any solubility, oxygen-binding affinity, or other property that we need, we might even see compounds like this in use one day colonizing new planets. The tenuous Martian atmosphere, for example, has a very small amount of oxygen in it naturally. Now, solutions of biomimetic oxygen carriers might someday be used to extract that small amount of oxygen from outside of a structure, pumping it back inside the building, where it's released, essentially, creating a building which breathes so that its occupants can too.

But our ability to imitate life goes far beyond creating small mimetic molecules. Remember that Kent was able to construct not only working copies of HIV-protease from L amino acids, but from D amino acids as well. This shows that even large biomolecules can be created abiologically and then used to enhance, or even create, biological systems.

Yet another exciting advancement was made by scientists at the J. Craig Venter Institute in 2010. This is when they reported that they had chemically synthesized a genome of over one million base pairs, then substituted that DNA for the native DNA in a bacteria. In this proof-of-concept experiment, the researchers made just a few small changes between the natural DNA of the bacteria and their synthetic form, targeting regions of the genome which were known to act as structural supports, rather than those coding for specific proteins. Now, such a simple change was enough that it could be detected in the cells, which had accepted the transplant, without compromising the viability of those cells. So the genetic deck was stacked in favor of success in this experiment.

Still, this achievement will no doubt take us to the next level of biomolecular engineering as it demonstrates clearly and effectively that the molecular machinery of life is modular and can be transplanted from one cell to another. This experiment shows that with enough understanding and cautious, dedicated effort, genetic material and the cellular machinery which uses it can be mixed, matched, and altered to produce any biochemistry that we desire. Someday, we may even create entire cells from nothing more than chemicals on a lab bench using organic reactions which are available to us today.

Just imagine what these cells might be designed to accomplish. We could literally synthesize self-replicating microscopic production plants for fixed nitrogen with nitrogenous activity manifold more powerful than rhizobia. We could create super red blood cells to enhance the oxygen-carrying capacity of blood used in transfusions. Imagine vast pools of photosynthetic bacteria of our own design harvesting light and converting it into biomass, or even energy, faster and better than any living thing ever has, all to the benefit of mankind. The possibilities are truly limitless. And the inspiration for it all hearkens back to that day in 1828 when Wohler accidentally created urea on his lab bench and realized that mastering the chemistry of carbon didn't require any mystical vital force, but careful study and application of the scientific method.

I'd also like to challenge you with one of humanity's major environmental concerns. We spend a lot of time looking at how chemists build upon carbon scaffolds to make organic compounds with new structures and properties. And we saw how some of the major entry points for carbon into organic chemistry include the hydrocarbons isolated from fossil fuels, which can be halogenated to make reactive substrates for substitution and elimination reactions.

We discussed how biological systems can also bring carbon into the organic world. Plants convert carbon dioxide into biomass, which makes its way through the food chain, where it's used by organisms to produce all the necessary chemicals to exist, some of which are extracted from those living systems by opportunistic chemists, like Justus von Liebig. But what about the exit point for carbon from organic chemistry? It's a fair question. It's actually a very good question.

As our activities as humans concentrates carbon in our fuels and biomass, we have to ask the question, where does that carbon go when we're done with it? Carbon-containing compounds contain a great deal of chemical energy. And whether we are burning fuels or using foods for respiration, a large amount of the organic material in the world is destined to become carbon dioxide. And although it isn't particularly pleasant to think about, when humans and other creatures are done with their biomass for good, decomposition naturally releases much of their carbon as CO_2 as well.

Now, that might not sound like a big problem. After all, carbon dioxide's a gas, so it's just carried away by the atmosphere, right? Well, not quite. We know from our discussions on spectroscopy just how powerful an IR absorber the carbon-oxygen double bond can be, because the vibrations of those very polar bonds in a molecule give it a strong absorption.

Recall also, from the Sir William Herschel's experiments that when objects absorb infrared light, they tend to heat up. So carbon dioxide acts as what we call a greenhouse gas, absorbing radiated heat from the Earth's surface, leading to climate change when it's not properly balanced. This has led to a global movement to devise ways to store and use carbon dioxide in ways which prevent its release into the atmosphere. For example, the U.S. Department of Energy has estimated at 2.4 billion metric tons of industrially produced CO_2 could be stored by injecting it into subsurface structures, like unmineable coal seams, where it can become absorbed to the surface of the underground deposit.

The concern here is that there are other gases already bound to coal in seams like these, most notably methane. Obviously, those who advocate this method are betting on irreversible strong binding of carbon dioxide to subsurface structures, like the carbon-rich coal deposits of coal seams, binding so strong that hopefully it can displace methane. Now, could such an idea work? Well, there's a good chance that it would. But it depends on how well each environmental gas absorbs to the stationary coal face relative to its tendency to vaporize. In complex geological systems like these, the overall composition and chemical behavior of formations can be tricky to predict. So we don't know the best method for sequestering CO_2 until we try and see the results. Just remember, it's all about those partitioning coefficients.

So we've considered quite a few ways that organic chemistry may change the world we live in. But it may also play a pivotal role in getting us off that world someday very soon. There are those who estimate that the Earth's current population of seven billion people would not be naturally sustainable;

that without chemical advancements like the Haber-Bosch process for nitrogen fixation, global famine would have already ensued, thinning the population to a number that the Earth could sustain herself. Yet, the global population continues to expand, pressing ever closer to the nine billion or so that could be fed using modern technologies. But what then? When we've completely exhausted the ability of our planet to provide for us, even with some assistance from science, what's next?

Even if we one day master the chemistry of nitrogen fixation, there will still come a day when there simply isn't enough land to feed every mouth on the planet. The natural next step is that we look beyond the Earth for places to live, continue to expand our knowledge and carry our species into the reaches of the solar system, and eventually the cosmos. To do so, at least to get started, will require that we take with us everything and everyone that a colony might need to survive transport to and establishing a colony on a new celestial body. But that leads to a number of logistical questions, not the least of which is, how do we get that much material off the Earth into space? It's a valid question. With current rocket technology, it costs thousands, or even tens of thousands, of U.S. dollars to get just one pound of material into orbit around the Earth.

So step one of colonizing space is to find a cost effective way to escape the gravity of Mother Earth. There are many possible approaches to this challenge. One very interesting proposal is the construction of a so-called space elevator, which would consist of a line which reaches from the equator of the Earth to an altitude of about 36,000 kilometers, the altitude at which geostationary orbit is obtained. Of course, building a tower more than 40,000 times the height of the Burj Khalifa presents some obvious structural challenges. It's unlikely that any such structure could support its own weight.

One proposed solution to this is to use a structure which is supported not by compression, but by tension, constructing a cable with a counterweight reaching so far into space that its center of mass is above 36,000 kilometers, and so its overall length is about 100,000 kilometers. This way, it will be held taut under its own momentum as it turns with the Earth. Now, using a cable like this, a mechanical climber could simply move up the cable, taking its payload of people and supplies to the necessary altitude to escape Earth's gravity with ease and reach outer space. Estimates are that a device like this, though profoundly expensive to build, once operating, could lower the cost of delivering material into space by more than tenfold. So it would cost just a few hundred U.S. dollars per pound to reach orbit. instead of the thousands, or even tens of thousands, of dollars currently required.

So in order to construct this modern marvel, a cable about 100,000 kilometers in length would have to be produced. Let's think about that for a moment. That is more than twice the circumference of the Earth, nearly a quarter of the way to the moon. But to build a cable that long which can withstand the tension produced by its orbiting counterweight, we have to find the lightest, most durable material with the greatest tensile-strength-to-weight ratio ever developed. And the front runner for that material is carbon nanotubes.

Under one current proposal, nanotubes several meters in length could be woven together into a rope, creating the longest man-made object in human history and providing us with the cable needed to make the space elevator a reality. Now, that three-meter length may seem like an insurmountable target for one molecule, especially considering that that rivals the length of a completely extended strand of human DNA. But researchers have reported that they have managed to create single nanotubes with lengths approaching the necessary mark needed to form that proposed cable. So it would seem that soon the only thing standing in the way of a space elevator will be our own imagination—and about $20 billion. If this project ever takes place, it would be a fitting application for these allotropes of carbon. The same element which makes up the scaffold of the molecules of life on Earth will also make up the scaffold of the vehicle which we will use to propel life forward into outer space.

So our daydreaming about the role of organic chemistry and surmounting the challenges faced by humanity and coming generations is coming to an end. But just because our time together is up, doesn't mean that your time with organic chemistry has to be over. If you found the last 36 lectures gratifying, I encourage you to carry your knowledge forward and use it as a basis on which to learn from other chemists. You now have all the tools necessary to read and digest current chemical literature created by researchers in the field, intended for those with an interest in the science of organic chemistry.

Countless university and public libraries subscribe to such prestigious professional journals as *Science, Nature,* and the Journal of the *American Chemical Society,* as well as more focused organic chemistry journals like the *Journal of Organic Chemistry* and *Tetrahedron.* I strongly advise you to search out these publications and others like them. They're a window into the incredible accomplishments being achieved every day by those toiling in the lab hoping to develop the next great catalyst, likes Sabatier, or find the new way to isolate vital compounds like Tswett. Perhaps they want to mathematically decrypt the kinetics of reactions, like Ingold, or decipher the function of a critical biomolecule, the way McKinnon did. And you now have the tools you need to join in on the conversation.

It has been my distinct privilege to help you get started on what I hope will be a lifelong journey exploring and appreciating the rich and beautiful chemistry which breathes life into our planet, the chemistry of molecules based on carbon. Thank you for joining me.

Glossary

acetal: The product formed from the reaction between an aldehyde and an alcohol under acidic conditions.

acid: A substance yielding hydrogen ions (H^+) when dissolved in water.

addition reaction: A reaction in which two molecules combine to form one product molecule.

aglycone: The newly added group on a glycoside.

alcohol: Organic compounds containing a hydroxyl (OH) group.

aldehyde: A compound of the type R(C=O)H.

aldose: A carbohydrate in which the carbonyl is placed at the end of the chain.

aliphatic: A compound that is nonaromatic.

alkane: A hydrocarbon in which all carbons are sp^3 hybridized.

alkene: An unsaturated hydrocarbon containing a double bond.

alkyl halide: A compound of the type RX, where X is a halogen substituent.

alkyl shift: The movement of an alkyl substituent from one atom to another.

alkyne: An unsaturated hydrocarbon containing a triple bond.

allotropes: Two or more forms of the same element, differing significantly in chemical/physical properties.

alpha helix: A secondary structure of proteins coiled like a spring.

amide bond: Also referred to as a peptide bond; a covalent chemical bond formed between two molecules when the carboxyl group of one molecule reacts with the amino group of another.

amine: A compound with nitrogen taking an sp^3 hybridization state, forming three single bonds to distinct partners: ($-NH_2$), ($-NHR$), or (NR_2).

amino acid: A molecule containing a carboxylic acid motif connected to an amine group via at least one intervening carbon.

amphiprotic: A molecule/substance that can donate or accept a proton.

amphoteric: A species that can act as both an acid and a base.

anabolism: The building up of complex organs and tissues in the body.

angle strain: Strain associated with the distortion of bond angles.

anion: An ion in which electrons outnumber protons to afford a net negative charge.

annulene: Cyclic molecules with alternating double bonds.

aprotic solvent: Any type of solvent lacking acidic hydrogens.

aromaticity: Special stability related with aromatic compounds.

base: A substance yielding hydroxide ions (OH^-) when dissolved in water.

Beer's law: $A = \varepsilon bc$, where A is the absorbance, ε is the sample extinction coefficient, b is the path length, and c is the sample concentration.

benzene ring: A ring of six carbons sharing double bonds.

bicyclic alkane: Two hydrocarbon rings sharing one or more atoms.

biomimetic chemistry: The practice of imitating biological molecules.

birefringence: The ability to refract light at two different angles, depending on the orientation of the electromagnetic waves making up the incident light.

carbocation: A positively charged hydrocarbon species.

carbonyl group: A group consisting of a carbon double bonded to an oxygen (C=O).

cation: An ion in which protons outnumber electrons to afford a net positive charge.

chirality: The existence of handedness of molecules.

cholesterol: The most abundant steroid in animals and a common starting material in biological synthesis.

chromatography: A technique in which organic compounds may be separated from one another based on one or more properties.

coding DNA: The portions of DNA that code for proteins.

conjugation: A phenomenon in which multiple pi bonds are in resonance with one another, lending extra stability to a compound or an ion.

continuous wave NMR: The most basic form/technique of NMR spectroscopy.

copolymer: Polymers constructed from multiple monomers.

covalent bonding: The sharing of electrons between two or more atoms.

crystal: Highly ordered, repeating arrangements of atoms or molecules.

cuvette: A cell, often made of quartz, used to hold liquid samples.

cyclic alkane: A cyclic chain of hydrocarbons.

Dalton's law: The sum of all partial pressures in a system is equal to the total pressure of the system.

deprotonation: The removal of a proton.

dewar: A vacuum-walled storage vessel used to hold liquids below ambient temperature.

diastereomer: One of two compounds that are non-superimposable non–mirror images.

Diels-Alder reaction: A reaction to make a six-membered ring by a [4+2] cycloaddition.

diene: A hydrocarbon containing two carbon-carbon double bonds.

dihedral angle: The angle between two substituent bonds on adjacent carbons.

dimer: A molecule formed by combining two identical molecules.

dipole: An intermittent charge buildup.

double bond: A chemical bond involving the sharing of four electrons between two atoms.

E1 reaction: An elimination reaction in which a weak base deprotonates a carbocation to give an alkene.

E2 reaction: An elimination reaction in which a strong base abstracts a proton on a carbon adjacent to the leaving group and the leaving group leaves, producing an alkene.

electrophile: An electron-deficient nucleus capable of receiving electrons from a nucleophile.

elimination reaction: A reaction in which a halide ion leaves another atom/ion.

enantiomer: One of two compounds that are non-superimposable mirror images.

enantiopure: A sample having all of its molecules having the same chirality sense.

enthalpy: The heat content of a substance, denoted by *H*.

entropy: Randomness; disorder.

equilibrium: A dynamic process in which the forward and reverse reaction rates are equal.

equilibrium constant: A constant (*K*) expressing the concentrations of both reactants and products at equilibrium.

ester: A compound of the type RO(C=O)R′.

ether: Organic compounds composed of two alkyl groups bonded to an oxygen (R-O-R′).

eutectic point: The only point/temperature at which a mixture of the two components can coexist at one distinct temperature.

fingerprint region: The lower-frequency region of the infrared spectrum containing a complex set of vibrations within the skeleton of a molecule.

Fischer esterification: A process of producing esters from an organic acid and an alcohol.

Fischer projection: A method of projection useful for conveying the arrangements of larger chains of atoms.

frontier molecular orbital: Orbitals that are involved in a chemical reaction, generally involving the HOMO of one reactant and the LUMO of the other reactant.

Frost circle: A method giving the Hückel pi molecular orbitals for cyclic conjugated molecules.

functional group: An atom or group of atoms in a molecule responsible for its reactivity under a given set of reaction conditions.

functional group region: The higher-frequency region of the infrared spectrum used to catalog the presence or absence of functional groups.

Gabriel synthesis: A reaction transforming primary alkyl halides into primary amines.

glycoside: A carbohydrate derivative in which the anomeric C1 group has been replaced by another group.

Grignard reaction: An organometallic reaction in which an alkyl/aryl halide adds to a carbonyl group of an aldehyde or a ketone, forming a C-C bond.

Haber-Bosch process: A process of converting nitrogen gas to ammonia.

halogen: Elements (fluorine, chlorine, bromine, iodine, astatine) found in group 17 of the periodic table.

halogenation: The replacement of a hydrogen with a halogen.

Hayworth projection: A representation of sugars in their cyclic forms with a three-dimensional perspective.

heteroatom: Any atom other than carbon and hydrogen.

HOMO: Highest occupied molecular orbital.

homopolymer: A polymer formed from one monomer.

hybrid orbital: A mixture of atomic orbitals (e.g. orbitals with both *s* and *p* character).

hydride shift: The movement of a hydride (hydrogen with a pair of electrons) from one atom to another.

hydrocarbon: An organic compound consisting solely of hydrogen and carbon.

hydrohalogenation: Reactions that produce alkyl halides from alkenes.

hydroxyl group: Also known as an OH group.

hyperconjugation: The delocalization of electrons in a sigma bond through a system of overlapping orbitals.

ideal gas law: $PV = nRT$; can be rearranged to demonstrate that partial pressures of gasses in a mixture are proportional to the number of moles of each compound present.

imine: A compound with nitrogen taking an sp^2 hybridization state, forming a double bond and a single bond: $R_2C=NR'$.

immiscible liquids: Liquids incapable of mixing together to form one homogenous substance.

internal alkene/alkyne: An alkene/alkyne having a localized pi system in the middle of a carbon chain.

ion: A charged species in which the number of electrons does not equal the number of protons.

ionic bonding: A type of bonding in which atoms of significantly different electronegativities come together.

isotope: Atoms that have the same number of protons but different numbers of neutrons.

junk DNA: The structural support regions of DNA that hold the coding pieces of the DNA sequence in place.

ketal: The product formed from the reaction between a ketone and an alcohol under acidic conditions.

ketone: A compound of the type $R(C=O)R'$.

ketose: A carbohydrate in which the carbonyl is placed in the middle/interior of the chain.

l-amino acid: An amino acid with the amino group (NH_2) on the left in the Fischer projection.

Larmor frequency: The frequency of precession of the nucleus in question.

ligand: An atom or group attached to another atom (in this case, a metal).

London dispersion forces: Attractive forces arising as a result of temporary dipoles induced in the atoms or molecules.

lone-pair electrons: Electron pairs not involved in bonding.

LUMO: Lowest unfilled molecular orbital.

magnetic anisotropy: The tendency of electrons in pi systems to be held less tightly and move through a greater volume of space than their sigma counterparts, creating a magnetic field of their own.

magnetic coupling: Synonymous with spin-spin coupling; a mechanism through which magnetization from one neighboring proton is encoded on another.

Markovnikov's rule: The addition of hydrogen halides to alkenes will occur with the halide to the more-substituted carbon.

mass spectrometry: A technique used to measure the molecular weights of atoms and molecules.

Merrifield synthesis: Also known as solid phase peptide synthesis; a method for synthesizing peptides and proteins with exotic side chains.

meso compound: Achiral compounds possessing chirality centers.

miscible liquids: Liquids capable of mixing together to form one homogenous substance.

mole: A quantity used to count atoms/molecules.

monochromator: A device used to disperse different wavelengths of light to the sample.

Newman projection: A method of projection useful for depicting dihedral bond angles.

nitrate: A polyatomic ion with the formula NO_3^-.

nitrile: Compounds containing a terminal nitrogen triple-bonded to a carbon.

nuclear magnetic resonance (NMR): In chemistry, a technique used for the structure determination of organic molecules.

nucleic acid: A substance found in living cells consisting of many nucleotides in the nuclei of cells.

nucleophile: A species that easily donates an electron pair to form a new chemical bond with another nucleus.

octet rule: An atom (other than hydrogen) tends to form bonds until surrounded by eight valence electrons.

optical activity: The ability of chiral molecules to rotate plane-polarized light.

orbital hybridization: The combination of atomic orbitals to produce hybrid orbitals.

organic acid: An organic compound having acidic properties.

organic chemistry: The study of carbon-based molecules.

organometallic chemistry: The study of compounds containing carbon-metal bonds.

***p* orbital**: A dumbbell-shaped orbital.

partitioning coefficient: The ratio of concentrations of a solute that is distributed between two immiscible solvents at equilibrium.

peptide bond: An amide bond formed between two molecules.

photon: A particle of light.

pi bond: A class of covalent bonds involving the overlap of *p* atomic orbitals in a side-to-side fashion, forming electron density above and below the bonded atoms.

pi system: Long, interconnected systems of *p* orbitals.

pi-to-pi* transition: The process in which electrons move from the highest occupied pi molecular orbital to the lowest unfilled pi molecular orbital.

pK_a value: A measure of an acid's strength.

plane-polarized light: Light that is composed of waves vibrating in one plane.

polarimeter: An instrument used to measure optical activity.

polymer: Very large molecules consisting of repeating units of monomers.

polymerase chain reaction (PCR): A process used to induce a DNA sample to replicate.

prodrug: A modified active compound that is later chemically converted into its active form by the patient's own biochemistry.

protein: A compound containing two or more amino acids linked together by amide bonds.

protic solvent: Any type of solvent containing acidic hydrogens.

proton transfer: The transferring of a proton from one species to another.

protonation: The addition of a proton.

pulsed NMR: Also known as Fourier transform (FT) NMR; often used as a more efficient NMR technique to acquire spectra simultaneously.

purine base: A nine-membered double ring with four nitrogens and five carbons.

pyrimidine base: A six-membered ring containing two nitrogens and four carbons.

quanta: A packet of energy that can only be absorbed in a transition of equal energy.

racemic mixture: A mixture containing both enantiomers in equal amounts.

radical: A species with unpaired electrons.

Raoult's law: A component in a mixture of miscible liquids will exert a vapor pressure equal to that of the pure solvent times its mole fraction in the mixture.

rate-determining step: The slowest step of a multistep mechanism.

regioisomer: Isomers that have the same molecular formula but differ in connectivity.

regiospecificity: A reaction that tends to occur at one particular position among other similar positions.

resonance hybrid: An average structure of all resonance contributors.

resonance structures: A collection of Lewis structures that can be drawn and differ only by the placement of valence electrons.

restriction enzyme: An enzyme that cuts the amplified DNA at very specific points.

retrosynthetic analysis: A technique used in which the synthetic problem is worked backward to minimize the number of reactions considered.

ring strain: The overall measure of the instability of a ring.

s **orbital**: A spherically symmetrical orbital.

saturated hydrocarbon: A hydrocarbon containing only sigma-bonded hydrogen and carbon atoms.

sigma bond: A class of covalent bonds involving the overlap of orbitals along the internuclear axis.

S_N1 **reaction**: A type of substitution reaction in which one molecule is involved in the transition state of the rate-limiting step.

S_N2 **reaction**: A type of substitution reaction in which the rate-limiting step involves the collision of two molecules.

spectroscopy: The observation of the interaction of light with matter.

spin: Also known as spin quantum number; usually refers to atomic nuclei having two possible magnetic states.

spin-spin coupling: Synonymous with magnetic coupling; a mechanism through which magnetization from one neighboring proton is encoded on another.

stereochemistry: The study of the three-dimensional structure and arrangement of molecules.

stereoisomer: Isomers having the same connectivity but different arrangements in space.

substituent: A different or more complex group in place that would otherwise be bonded to a hydrogen.

substitution reaction: A reaction in which an atom is replaced by another atom.

substrate: The entire molecule undergoing substitution.

tetrahedral: A molecular geometry consisting of an atom at the center connected with four other atoms, creating a tetrahedron with bond angles of 109.5°.

torsional strain: Strain associated with the resistance to bond twisting.

triple bond: A chemical bond involving the sharing of six electrons between two atoms.

unsaturated hydrocarbon: A hydrocarbon containing pi bonds.

valence shell: The outermost shell of an atom that is comprised of valence electrons.

Williamson ether synthesis: An organic reaction used to form an ether from an alkyl halide and an alcohol.

X-ray crystallography: A very powerful tool used for structure determination of crystals.

Zaitsev's rule: The major product of an elimination reaction will be the most-substituted alkene possible.

Zeeman splitting energy: The energy difference between the alpha and beta states.

zwitterion: A molecule/species containing a positive and negative charge in separate regions.

Bibliography

Berson, J. A. *Chemical Creativity: Ideas from the Work of Woodward, Huckel, Meerwein, and Others.* Weinheim: Wiley, 1999. A brief review of the contributions of some of the more recognizable names from the first half of the 20th century. Technical and detailed, this work cites many of the original research papers written by the scientists themselves.

Bulletin for the History of Chemistry. A publication of the American Chemical Society's Division of the History of Chemistry. All but the most recent three years are available free of charge online at http://www.scs.illinois.edu/~mainzv/HIST/bulletin/.

Carson, R. L. *Silent Spring.* New York: Houghton Mifflin Harcourt Publishing Co., 1962. The seminal work in environmentalism published during a heyday of commercial chemical synthesis, Carson's work was the first to seriously raise concerns over how this explosion of new chemicals released into the environment might have unintended consequences.

Cobb, C., and Harold Goldwhite. *Creations of Fire: The Path from Alchemy to the Periodic Table.* New York: Plenum Press, 1995. A wandering compendium of anecdotes and facts about nearly every influential chemical thinker in the last 2000 years. This will prove to be a difficult read from start to finish, but it is better used in the spirit of an encyclopedia of great researchers, their contributions to chemistry, and their connections to one another.

Deamer, David, and Jack W. Szostak. *The Origins of Life.* New York: Cold Spring Harbor Laboratory Press, 2010. This is a compilation of perspectives on the chemical origins of life written by researchers in the field.

Edwards, B. C., and Eric A. Westling. *The Space Elevator: A Revolutionary Earth-to-Space Transportation System.* BC Edwards, 2003. A book detailing the engineering and technical challenges still facing a hypothetical space elevator, written by the author of one of the definitive studies on its feasibility.

French, J. *The Art of Distillation: An Alchemical Manuscript Being Certain Select Treatises on Alchemy and Hermetic Medicine.* Calgary, Alberta, Canada: Theophania Publishing, 2011. One of the recent reprints of the 1650 classic by John French. This work offers a fascinating insight into how thought was changing during French's time as alchemy slowly gave way to chemistry.

Hager, T. *The Alchemy of Air: A Jewish Genius, a Doomed Tycoon, and the Scientific Discovery That Fed the World but Fueled the Rise of Hitler.* New York: Random House, 2008. Hager masterfully weaves the history of western Europe in the early 1900s with just a touch of chemistry to bring context to Haber's unrelenting quest to fix nitrogen from the air. Some of Haber's other chemical endeavors, such as chemical weapons and an attempt to harvest gold from seawater, are also discussed.

McMurry, J. *Fundamentals of Organic Chemistry.* 7th ed. Belmont: Brooks/ Cole, 2011. A less-extensive introductory-level text than Wade's offering; some readers might find this text easier to navigate.

Morris, R. *The Last Sorcerers: The Path from Alchemy to the Periodic Table.* Washington DC: Joseph Henry Press, 2003. This book surveys humanity's understanding of elements over the ages, from the first recorded suggestion that such materials exist all the way to the discovery of subatomic particles.

Scott, Raymond P. W. *Techniques and Practice of Chromatography.* New York: Marcel Dekker Inc., 1995. This is a thorough manual describing most modern chromatography techniques, rooting them in historical context.

Spectral Database for Organic Compounds. http://sdbs.db.aist.go.jp. Accessed July 9, 2014. This free online database of molecular spectra is compiled and managed by Japan's National Institute of Advanced Industrial Science and Technology (AIST). Many of the representative spectra in the course are adapted (with permission) from the spectra available on this site. It is a wonderful place to explore spectra and hone your spectral interpretation skills.

Stephens, Trent, and Rock Brynner. *Dark Remedy: The Impact of Thalidomide and Its Revival as a Vital Medicine*. Cambridge: Perseus Publishing, 2001. This book surveys the impact of thalidomide on the history of the world. Written by a professor of anatomy and embryology, this is a riveting narrative of how blind faith in the power of organic chemistry can have disastrous consequences.

Wade, L. G. *Organic Chemistry*. 7th ed. Upper Saddle River: Prentice Hall, 2010. An excellent, comprehensive introductory organic chemistry text. It is currently in its 8th edition, but many 7th-edition copies are still circulating.

Walker, S. M. *Blizzard of Glass: The Halifax Explosion of 1917*. New York: Henry Holt and Company, 2011. A historical account of the events leading up to, during, and after the legendary Halifax explosion of 1917. Although this is not a chemistry text, it will leave the reader with a sound impression of the immense energy contained in some organic compounds like picric acid.

Watson, J. D. *DNA: The Secret of Life*. New York: Knopf, 2004. This is a historical account of the impact of DNA research on society written by one of the discoverers of the double helix, James Watson.

Williamson, Kenneth. *Organic Experiments*. 9th ed. Boston: Houghton Mifflin Company, 2004. This university-level laboratory text offers detailed explanations of techniques used for synthesis and purification of organic compounds.

Notes